P9-DDD-379

The Names of Things

THE NAMES OF THINGS

A Passage in the Egyptian Desert

SUSAN BRIND MORROW

RIVERHEAD BOOKS
A member of Penguin Putnam Inc.
New York • 1997

For Lanny

Riverhead Books
a member of
Penguin Putnam Inc.
200 Madison Avenue
New York, NY 10016

Published simultaneously in Canada

Library of Congress Cataloging-in-Publication Data

Morrow, Susan Brind.
The names of things : a passage in the Egyptian desert /
by Susan Brind Morrow.
p. cm.
ISBN 1-57322-027-2
1. Morrow, Susan Brind. 2. New York (State)—Biography.
3. Egypt—Description and travel. 4. Americans—Egypt—Biography.
5. Deserts—Egypt.
I. Title.
CT275.M6465A3 1997 96-44505 CIP
973.92'092—dc21
[B]

Printed in the United States of America
10 9 8 7 6 5 4 3 2 1

This book is printed on acid-free paper. ♾

Book design by Debbie Glasserman

CONTENTS

You could keep some remnant of it, a talisman that would become rare and fine, worn over time into something familiar. It would naturally become more thin and precious the more the air wore it out, like the bones of a saint. After all, it was only an object in the physical world, not something more potent, like something in the mind: memory.

But the original, the thing itself, would never come back. It had passed away from the world. You could conjure it, though, the emotion that kept it alive inside you, with a trigger: an image, a smell, a combination of sounds that formed it into a picture that stayed in your mind. That was the life of the thing after it died. The only thing that could bring it back.

This is what a word is worth.

DEAD LANGUAGE

YOU COULD BEGIN with the crab that scratches in the sand. The name of the animal is the action or sound it makes, or its color. The name parents other like meanings belonging to other things, leaving the animal behind: grapho (Greek—to scratch, and so, to write), gramma (the scratches), graph, grammar, grab.

As I walk along the shore of the Red Sea at dawn a hundred pale pink crabs scuttle carefully back across and into the white sand. Behind a sharp crust of coral a rock crab, seaweed-green edged with red, pries the back off a sand crab and feeds. It is not so easily frightened and merely watches me. There are tiny porcelain-blue crabs in the mangroves a few miles south, popping out of the dense muddy quicksand like living jewels.

In this harsh environment life itself is a gorgeous miracle, coming out of the barren desert, out of the bitter sea: hals, the sea of salt.

Above the tide line the sand is crusted over with glass, hard-surfaced and brittle like frosting sugar. It snaps into square panes of rock. Rocks flecked electric blue-green with what became copper here wash down in the mountain floods. Walking in these hills I am looking at visual puns. I can see how readily the creatures translate, were translated long ago, into thought and use and language. What is lost is a sense of their intense beauty: that they are alive.

Words begin as description. They are prismatic, vehicles of hidden, deeper shades of thought. You can hold them up at different angles until the light bursts through in an unexpected color. The word carries the living thing concealed across millennia.

"The lynx is speckled like the starry sky," the nomads here say. "The crane belongs to the rain, and the stork belongs to the sea."

Animals belong to their environments, are inseparable from the processes that draw them. As Saad told me, pointing down to our feet where qatta, desert grouse, were moving over the stony ground. I could not see them for some minutes, so much did they resemble the brown-black scatter of stones. "You see," he said, "as you are white like snow and ice where you come from, and I am brown like rock and sand. Don't write that down."

He tells me of the mystery of domesticated goats, why they are white and black, unlike their cousins, their ancestors, the ibex, who are invisible against the rock cliffs they climb. An animal taken from its environment may lose its natural color. Color is a defining principle of place.

The flamingo is the hieroglyph for red. All red things: anger, blood, the desert are spelled with the flamingo. The Red Sea Hills are mostly red. The red rock is vibrant in the changing light.

Near here are lavender mountains with cranberry cliffs. Silver and blue and green wadis wind around them. But the true red of the Eastern Desert, the red of Wadi Baramia, of Nugrus, is an intense color, harboring little plant life except the sweet-smelling selim that grows straight up in branches from the ground making the best walking sticks. It is a painful color, harsh to the eyes.

Flamingo, flaming. In Greek its name is phoenicopteros, phoenix, with feathers of fire. The riddle extends: the delicate, breakable flamingos breed on ash cones in the evaporate bed of Lake Natron in Central Africa. The new birds arise from the ashes. Fee waqt el mattar, in the time of rain, they arrive in the thousands. Last February there were twenty thousand flamingos on Lake Bardawil in North Sinai.

Red and green define the environmental extremes of Egypt. The desert is red. The sea is the great green. The sweet sea, the Nile, was once clotted with papyrus, thriving, gigantic, mobile, filled with animal and bird life, as it is today only in the Sudd, the great marsh in South Sudan. In Egypt the plant no longer exists. It survives only in the hieroglyph for green.

Rock drawings are scratched onto the flat, wind-scraped surfaces of red sandstone. They illustrate how the landscape has changed. The oldest are of giraffe, elephant, rhinoceros, from when this bare rock was grassland.

These African animals are among the first hieroglyphs. The giraffe is in the verb to foresee. The saddlebill stork, now rarely seen north of Khartoum, is the picture that defines the word for soul.

These early written words emerged as the land here rapidly became desert in the Neolithic and pushed what they represented away, as though the growing desert isolated and so emphasized each living thing.

In ancient Egypt hieroglyphs were called medew netcher, sacred words. Netcher is the picture of a flag on a pole, like the flags that mark sacred places throughout the desert even now:

tombs and rocks and trees. It eventually became the Coptic, hence early Christian, word for God.

Five thousand years away there is an incidental pun: netcher equals nature.

I come from a dislocated people, the Scots who came to Canada in the nineteenth century. I have never been to Scotland, but I have made the journey to Canada many times, the three-hour drive from my parents' house to Kingston, where my mother grew up. At least once a year we cross the slender green bridge over the St. Lawrence, through the rose-pink cliffs of Wellesley Island and into fields where lichen-spattered whalebacks of grey granite make the land poor.

David Hutchison, my father's grandfather, arrived here on the northern edge of Lake Ontario by steamboat at the age of five. His father had purchased two hundred acres north of Toronto, where the boy would learn to shoot and ride.

"They wanted to live like Scottish lairds!" My great-aunt Dorothy laughs as she describes them, coming to the wilderness with all their grand designs. Her own father, David Murray, my mother's grandfather, sailed from Scotland to Ontario around the same time. He saw his family lose almost everything to the infertile soil of the Canadian Shield, the oldest exposed rock in the world.

But these were Scots. They were tough, they got by.

David Hutchison went on to study theology and ethics with my mother's great-uncle John Murray, who had become the head of the philosophy department at McGill. As children we heard stories about my great-grandfather. He was, we were taught, that best of all possible things a person could become. He was a polymath.

He had taken degrees at Harvard (in medicine) and Berkeley (in law). And, my father would say, always talking about his namesake with great tenderness and amusement, he knew seven dead languages. This seemed an astonishing feat, and at the same time, as told by my father, it was clearly meant to be funny, an oxymoron given to a child as a toy. What after all was a dead language?

"Well, let me see . . . he knew Aramaic, Greek, Hebrew, Latin . . ."

Then he would forget what the others might have been and tell us instead about Rex, his grandfather's dog, who had one blue eye and one brown eye and fell into the goldfish pond. We knew the goldfish pond. We used to bring the fish into the basement in the wintertime, my sister Barbara and I, tossing their mottled bodies into metal tubs with minnow nets, and hauling the tubs down the back stairs of my grandparents' house in Albany.

And then there were the stories, much more exciting, about

David Hutchison as a traveling Presbyterian minister in Montana and British Columbia at the turn of the century. David Hutchison riding with the Crow.

I never knew my great-grandfather. I have only one old black-and-white photograph of him, from the late 1940s, when he was a professor of Hebrew (or was it at that point pharmacological law?) in Albany. He was near eighty at the time. In the picture he is sitting on a slatted bench by the fishpond, a white hydrangea bush close beside his head. His dress is antique, his boots, tie and suspenders, bunched white shirt, brown felt hat. He holds a box of goldfish food between his two hands. At first glance it looks like a book he is about to open. His face looks away, distracted. I know the face well. It is the shape of my father's.

My grandmother Laura spoke of her intense loneliness growing up. Of having to wear, she said, her mother's old evening gowns from Montreal cut down to fit her, because there was nothing out there in the frontier towns, no place to buy clothes, only the Indians for friends.

My grandmother died in a nursing home in Phoenix, Arizona, in 1981.

My father and I went to visit her there. One day she put her hand on mine and said, "Can you hear them? The Indians keening in the Catskills?"

"That's the air-conditioning, Mother," Aunt Nancy said.

—

"Netcher equals nature." A friend of mine said this to me once, early one morning, on a decaying boat in a dying river. We had been up drinking whiskey in the raw cold of that night. It was winter, the beginning of the new year, the new decade. There was a space heater on at our feet. Its vibration, or the wind outside, rattled the brass curtain rods against the window glass.

He knew something about Coptic. He had married a Copt. He had lived in Egypt for years, and when he went back to America, as a professor, an Egyptologist, he knew the country well. He had learned Egyptian colloquial, and understood the multiplicity of Arabic words. How a single triliteral word with its fluid, hidden vowels can blossom into many forms, can run backward and forward, can easily become, because of this ability to slide into another dimension of meaning, a joke or a curse. He knew that in Egypt a pun is not an accident. A name is a mirror to catch the soul of a thing, and a pun is the corner of its garment.

We sat through the night in the soup of disintegrating forms that is modern Cairo, modern Egypt. "How can we talk about netchers," he said, by which he meant "primary hieroglyphs," pristine archaic nouns, words that would be drawn directly from nature.

I had thought about them for years. My own name was one.

It was one of the oldest words in the language, I knew. A nonsense word, a reduplicative, a child's toy. A core word, the kind of simple word you would expect to find in the desert, where they lie about as polished and various as the colored stones on the scrubbed wadi floors.

You could imagine a child rolling it off of her tongue in play: susa, shusan, a common thing—as a primary hieroglyph would be. A name as common as the thing it was made to suggest, a water lily. And yet it was an important thing, because it was a marker; its appearance meant the presence of something not common enough in a desert country: water, sweet water and mud.

I have an object from this desert: an ostrich feather fan, the small inner grey feathers of an ostrich tied together with dime-store string. I press its soft fur against my face and remember the day in Wadi Kharit, the day I met Saad, the young Abadi who had killed an ostrich in a wadi near Kom Ombo. The first ostrich, we thought, to be seen that far north in years. The rains brought it, the heavy rains that caused the flooding in Sudan in 'eighty-eight and made the desert green and brought out of it the clouds of locusts later on that year. I sat in an edge of shade that scorching afternoon in November on a rug thrown down on the stony ground. I sat with Joe, who had brought me there, and Saad came and set down beside us the head of a small gazelle, its eyes frozen in blood, that he had tracked and slain that same day, and brought for us a sampling of its tender meat, its heart and liver on a white tin plate.

I spread the feather fan across my face and inhale it. Its smell is vague and fading now, like the scent in my journals from that time, the scent of some desert plant, sharp and lemonlike, or of the wild thyme or sage I once pressed in their pages—having torn a few leaves of the plant and crushed them to draw the scent into my hands, letting it relieve the stark light and the heat of the day as I walked.

When I thought of Africa as a child it was not of a desert. There were stacks of clothbound Tarzan books in my grandfather's attic with the smell of moist decay in them. The cheap old paper was yellow and chipping away like old paint. I read them in the summertime, stretched out over weeks of daydreaming. The Africa of my imagination was a dense green world filled with enormous trees, a darker, fuller version of the forest world I knew. Though Tarzan was not among the stories my brothers and sister and I acted out.

On those warm days we scattered through the woods and fields around our cottage in the Finger Lakes. Our house was the hedgerow, the hollow between the arching trees, Osage orange and chokecherry bounded by corn and wheat. We were avid for every detail of where we were: we dug fossils out of the chipped slate. We suffocated butterflies in clamshells, and caught snakes and fish and buried them and dug them up months later to see their bones.

The Finger Lakes, the low rolling glacial terrain south of Ontario where the God of the Iroquois had pressed his hand upon the earth.

Seneca is the largest lake, the middle finger. We lived on its north shore in the winter, in Geneva, and on the east in the summer, on East Lake Road. I remember the winters as hard. There were snow days, when we did not have school and when the sky was a rare pale blue, as though its lowest arc of light mirrored the deep snow.

The cold was enough to hurt my feet as I walked home from school on the slate sidewalks, and enough to curl our breath out white in the morning air. But the lake never froze, as Mrs. Mackey, my first-grade teacher, then in her eighties, said it used to do. In 1912 a thousand swans came down on the lake to land and froze to death. Their feet stuck to the ice and they could not take off again.

Seneca is the deepest of the Finger Lakes. A mysterious thud would occasionally quiver up from its bottom trough of sludge hundreds of feet down in the middle of the night: the phantom drums of the long-departed Seneca Indians. In its cold lower depths, we believed, corpses were preserved for years and great corpse-grey sturgeon lurked.

There were small clusters of cottages around the northern edges of the lake, but beyond them, for forty miles, the surrounding hills buckled up into farmland, woods, and vineyards with few visible roads or houses. Hidden in the woods on the east side of the lake was the nuclear arms depot. We knew it only for the deer that appeared along the inside of its high mesh fence in the evening to feed. They were white, and ghostlike, frozen in the headlights of our passing car.

Our first understanding of nature came in this way, in intense glimpses: my mother tracing out an owl in the black recesses of a neighbor's pine tree; waking me to show me the moon, to stand me in its pool of silver light on the floor; and one summer night taking us out into the field with a blanket to learn the stars: Vega, the Dippers, Cassiopeia webbed like a hand.

Fishing was a binding principle in my family. This was not glamour fishing. As my brother David once famously said, while addressing the Geneva Rotary Club when he was in high school, "Salmon's a trash fish, Mr. Marvin." (By which he meant, with all the grace and humor of his sixteen years, "We don't fish for salmon in the Finger Lakes. Salmon fishing belongs to the world of television, of tourism.") David was a tournament bass fisherman at the time.

The fish we went for were the fish of the northern lakes, deep green fish splashed with white like light through water: pike, pickerel, and in Canada and the St. Lawrence muskie and walleye.

My mother was Canadian, and the presence of Canada was always near. I have a strong early memory of being carried across a starry field on a cold night, wrapped in an old satin quilt, by Uncie, my great-uncle Ernie. Uncie had been shot down over Germany in World War I. Gangrene ate into his leg in a prison camp there, and when he came back to Canada after the war he never worked again. He married my great-aunt Dorothy and they lived in the woods, in old United Empire Loyalist houses that were cheap to rent, that were too big and cold for anyone else to want. They had no money whatsoever (I was being carried to an outhouse in the early 1960s). They lived in a cabin at Abbey Dawn, and my aunt, a poet-naturalist, gave readings there by the big cracked bell.

My aunt was deeply Christian, and yet it was clear to me even as a child that her God had something to do with a mystical sense of a pervasive holiness in nature, with the workings of the world as it is.

—

I remember her, as she so often was, poised and listening, and then saying in her deep Scots voice something about the loons on the lake, the redpolls down from the Arctic that year, the chickadees, their comings and goings. Something informed and clear, but full of presence, about rocks (the flecked pink granite of southern Ontario) or trees (the peeling-paper birches, the sumac, the white pines). She never saw the sea, and knew utterly, absolutely, where she was, every sound and smell.

Aunt Dorothy and Uncle Ernie in part raised my mother, whose father never recovered from the war and killed himself, in 1930, when she was three years old. And Aunt Dorothy in part raised me, in letters, in books she sent, in visits, holding out my hands filled with peanuts to feed the birds. She gave us some measure of this reverence of hers, and as children we lived it in our different ways.

My older sister Barbara went off into a neglected field thick with wild roses to perform her grave, secret rituals.

I rose in the hour before dawn. I would go out into the rich darkness, into a thrilling invisibility, and, intensifying it, swim in the blackness of the lake, in the path the moonlight made. I would walk east to the marshy ground where herons sat folded in the tall dead elms, and watch the first streaks of green swell with light and shrink and deepen into dawn. The sun came quickly, splashing everything with filmy gold, and then just as rapidly fading it out, paling the first vivid blues and yellows of the roadside flowers.

Just as I learned currents in the day, the qualities of the different hours, I learned the marked difference between the inside and the outside.

There was always a voice in those days calling me back inside, hinting at the dangers of this outside world. It grew louder and I ceased to trust the dark. I walked to the lake with a flashlight. The beam of light fell not on a broad blank surface of water, but across the backs of fat grey carp, poised and still in lines to spawn throughout our shallow bay. I fled.

BARBARA WAS KILLED in a car crash late one June night on East Lake Road. We knew every fold of that road. The driver did not, and, drunk, flew off into a grassy ditch.

Our house in Geneva filled with people from the town, people from the North Presbyterian Church, who came to do what they could to help our family through, to hold us together. A family is like a house, a structure in itself, the first structure in the psyche. The people in it are the solidest things you know, enclosing you with all the warmth and range of human possibility. The house of my family had begun to crumble. A wall had abruptly caved in. People came bringing food that no one could eat. They sat in the living room, where my parents were quietly weeping.

I withdrew into my blue-and-white second-story bedroom, closed the curtains and closed the door. I sat at my desk drinking cream sherry (there was a special glass for this ceremony: a champagne glass, the stem of which was an opaque, tousle-headed Dionysus) and did my Latin homework. I was fourteen years old.

My childhood was over. But the lake stayed with me. I see the dark blue strip of its water whenever I orient myself in the world. West is straight across, where the sun sets in August, with Scorpio rising to the south. The church spires and white Agway tower (long since burned down) of Geneva are the northern rim, and I stand in the east at the end of our dock. I automatically impose this template when I emerge from the subway in lower Manhattan, to figure out the east and west, the north and south, before I know which street to walk.

Two years later, when I was sixteen, I went away to college in New York. When I moved to the city my sense of being closed in, of a palpable greyness, became overwhelming. The streets were like caverns. I watched pigeons drop diagonally from the buildings and wheel suddenly together, imagining, but not knowing then that they were mountain birds.

When I went home to visit my parents I stayed in bed for days. It was a rainy fall. The smell of raw wet earth came through the open windows, and I grew to love it as though noticing it for the first time.

In my one-room apartment on 112th Street (having moved

out of the dormitory, having insisted on living alone), an old puff hair dryer of my mother's on my head to soften the jawache of TMJ, I sit at my corner table with Plato, the Oxford text, open before me. Jowett (translation and commentary) is open at one elbow, and Liddell and Scott (the dictionary) at the other, and I am looking up and writing down every single word.

"What a hopeless, one-sided way to learn a language!" my old friend Carl used to say. "You can't speak it. You can't even pick up a book and read it. You just look it up, word by word."

But that was the whole point, I tried to tell him, looking up the words, as though every one were an entity in itself, a picture. To read Greek is to know how to look at a word on a page, to examine a word long before you know what it means. You are looking at the bones of language. There are words, I would say, that have a tactile quality, like phrix, the stiffening of water in the wind, or the skin in fear.

Such a word belongs to the vivid nature imagery of the early poets, with their desperation to pin a thing down with a metaphor, something alive: the shrill reed pipes that were the hissing of serpents on the abruptly severed, dying head of Medusa, or arose from Syrinx, the wood spirit transformed into a bed of rushes as Pan caught and crushed her, and the wind came shrieking through.

There was a book I particularly loved when I was a teenager,

David Campbell's *Greek Lyric Poetry*. It was a collection of fragments, many of them found in Egypt scrawled on the shreds of mummy wrappings from crocodile cemeteries, on the backs of discarded tax forms, or on broken pots in trash heaps—so scarce and valuable was paper—from cities rendered desert by inevitable drought, the land turning to salt.

"The papyrus was discovered in 1855 at Saqqara by the French Egyptologist, Mariette, and is now in the Louvre. . . ."

It was a small red book, my copy soon battered and unrecognizable, bandaged back together repeatedly over the years with black electrical tape. Its broken lines of intense descriptive poetry were like shards of some wonderful glass—a lens on the world, shattered into concentrated pieces.

Bale de bale kerulos eien
os t'epi kumatos anthos am' alkuonessi potetai
nedees etor exon, aliporphuros iaros ornis

Would oh would I were a kingfisher that flies with the halcyons along the breaking waves, with a fearless heart, that holy bird, the deep blue of the sea . . .

The birds and trees in these fragments were often untranslatable, not yet matched up with living things: the halcyon: "see

Thompson, *Greek Birds,* 46–51; 'symbolic or mystical bird, early identified with the kingfisher.' "

The kerulos, cerulean, a bird defined simply by its color. The kingfisher (as kerulos was translated in the dictionary) does not fly over breaking waves. It belongs to lakes and rivers. And yet what other bird in the Mediterranean is that deep a blue?

Carl would only laugh when I told him about what I was reading, about how Alcman learned meter by listening to the songs of birds. Carl was two years older than I was and sometimes hovered protectively around me with his friendship. As I sat in the Gold Rail with some doomed new infatuation, the popular music of the mid-seventies thumping through our talk, Carl was always there in the crowd waiting to walk me home.

We would sit on an icy bench in Riverside Park, draped in our long dark coats, and he would pull out of his pocket a piece of cake he had gotten especially for that moment. Now we would have . . . a feast of cake. We would forget about those Columbia boys, about the Gold Rail. He would then pry my notebook out of my cold hands and, looking too closely at what I had written on the page he opened to, would write in huge letters at the top:

DOES ENJOYMENT PLUS ENJOYMENT EQUAL MORE ENJOYMENT?

Then he would show me something he knew and I did not: how to make angels in the snow.

Carl understood the beautiful gesture. He could with a sudden flurry and with very small, ordinary props pull a moment of delight right out of the air.

Sometimes as we walked downtown along the park I sang to him a torch song I remembered from my Geneva years:

My story is much too sad to be told
But everything leaves me totally cold. . . .

It was Carl who first told me about the Red Sea, that this was where Mary Magdalene had gone to die. He had seen it in a painting, Mary Magdalene with her blond hair flowing into her hairshirt, the hair flecked with gold that was still bright after five hundred years. Carl had an interest in paintings from the fifteenth century. Often in them, a tiny greyish city in the background (for the story in the picture usually took place far away from the city and its rim of green fields, in a wilderness of sharp blue stones) would be made of crushed lapis lazuli, and real gold would be used here and there to denote something remarkable: a moment of grace, or the presence of a saint. Carl loved the idea of saints and angels, and eventually he became a scholar of their iconography, the clues to who they were.

There were other paintings of the desert on the Red Sea that we talked about—the series of Sienese paintings that illustrate the life of St. Anthony, the founder of Christian monasticism. These had a delightful cartoon quality, with the saint shown at two or three different stages (of a journey, coming down a staircase) in a single panel: St. Anthony gives away everything he owns, and goes to live in a cave on a mountain above the Red Sea. He has a dream that he is not alone, that there was an even older hermit in a cave on the other side of the mountain. St. Anthony, with his crooked stick and long white beard, sets out to find the aged St. Paul in his cave, and finds him just as he is about to die. As St. Anthony sets out on his journey to find St. Paul he walks through a dense forest. A centaur comes out from behind a tree and tells him where to go.

We thought this was very funny. What is a forest doing in the desert on the Red Sea? And a centaur? Another painting in the series gave a more accurate representation of the Red Sea, of the Galala Plateau, where the Monastery of St. Anthony is. I think of this painting in conjunction with the one that hangs beside it on the green velvet wall of the Lehman Collection in the Metropolitan Museum of Art in New York—the Giovanni di Paolo in which an angel tells Adam and Eve that they must leave the Garden of Eden. The angel approaches them with great urgency as they walk in delight beside a line of golden pear trees. There is something behind them that they have failed to notice, it seems, and the angel has come to point it out. It is a representation of the world they are to enter: a flat circle, like a mandala, ringed with exquisitely beautiful deepening blues. The outermost rim of blue is stamped with the golden signs of the zodiac, the circular swing of time. In the center is the actual landscape where they are to live. It is a harsh place, scorched and brown, with small isolated patches of green. Black rivers come down from a mountain of bare rock and thread through the desiccated land like the veins in a hand.

The St. Anthony painting beside it takes you immediately into this landscape: the saint walks along a stony path into the mountains above the Red Sea, which is a dismal pea-green, as though its water is in deep shadow. On either side of the path are small twisted leafless trees. Animals lie as still as stones beneath them. There is a faint pinkish tinge of red, which can only be—and you see this standing back and squinting—the sliding light of the low sun, a thin luminous veil. The painting is such an accurate account of afterglow, the absorption of light in the desert, that I often go to look at it now when I want to remember the place.

I had a teacher in those days named Steele Commager. Commager's familiar lurching form (squash racket tossed in a green

Harvard Coop sack over his left shoulder; little cigar, with its long high curl of yellow smoke, between his thin lips) emerged from time to time from the elms on Riverside Drive, where he lived. I would catch glimpses of the odd high twisting branches of these old trees down a steep side street as I walked the particularly bleak stretch of Broadway around Columbia.

Commager would come to class with a stack of loose yellow pages marked in pencil with his notes from years before, his long pale hands sometimes shaking a little as he ruffled through them. He did not seem to have much enthusiasm for teaching.

"The only thing I ever learned from Steele Commager," Carl used to say, "is not to light a gas stove in the afternoon when I'm wearing a tie. Which is actually very good advice."

And yet he loved words.

Commager taught us about the hapax legomenon, the word that is used only once, that is created for that occasion only. The hapax legomenon in my mind was a rare animal that could be tracked. I imagined it with its unusual form trailing through the dense forest of language. Commager was always having us track things, word origins, poetic lines.

I ran into him one Saturday morning in the catalogue room of Butler Library. This was a piece of luck. One rarely got the chance to run into Steele Commager. I rarely got the chance. I had been trying for months, searching the streets of Morningside

Heights at precisely odd times like this when the likelihood of running into him was dramatically diminished.

But there he was, in all his scratchy yellow radiance, rifling through the walls of old wooden drawers. Seeing him so suddenly, I froze, not sure whether to approach or flee. He looked up at me before I could decide what to do. He gave me his low, flickering smile. "Ah, Miss Brind," he said, "I'm looking for a line, 'Who bridled behemoth? Who curbed the wrath of leviathan?' "

I resolved to find it.

My father had an amused, particular fondness for oversized animals. I remember him bouncing me on his knee when I was perhaps three or four while he chanted a children's rhyme that ended with the word hip-po-pot-a-mus, and was a description of the animal. He loved the passage in Job (40.15):

> Behold now behemoth which I made with thee; he eateth grass as an ox. Lo now, his strength is in his loins, and his force is in the navel of his belly. He moveth his tail like a cedar: the sinews of his stones are wrapped together. His bones are as strong pieces of brass: his bones are like bars of iron. He is the chief of the ways of God: he that made him can make his sword to approach unto him. Surely the mountains bring him forth food, where all the beasts of the

field play. He lieth under the shady trees, in the covert of the reed and fens.

The shady trees cover him with their shadow; the willows of the brook compass him about. Behold he drinketh up a river and hasteth not: he trusteth that he can draw up Jordan into his mouth.

I told my father when I was at Barnard that behemoth was simply the Egyptian name for the animal, as hippopotamus was the Greek. You could see it easily in the hieroglyph: pa mu, water ox.

Later I was in Queen Elizabeth Park, an abandoned game park in the Ruwenzoris on the border between Uganda and Zaire. Idi Amin's soldiers and the Tanzanian army had killed most of the animals in the years before while they were fighting each other. They gunned down whatever they saw. I heard there was a lot of haphazard shooting from helicopters.

When I was there in the late spring of 1983, the only thing that seemed to remain were the marabou storks, and the pelicans and flamingos that fell from the sky in huge numbers, at regular intervals, like falling snow.

Around the muddy edges of Lake Albert, where the birds came down, were hundreds of hippopotami, half sunk in the dark watery mud. Their slippery pink mouths would open and shut, slowly, occasionally, like the dark upper flaps of old pianos exposing the square white chunks of their teeth.

The first time I saw a hippopotamus track I couldn't figure out what it was. It looked as if the post of a small house had been pried from its foundation deep in the ground. I walked around the game park by myself and began to see these tracks everywhere, along the green grassy slope that rose up a sharp hill from the lake. I slept in a yellow concrete shack that had no door, no windows. The soldiers had taken everything. I began to understand what I could plainly see. The animals came out of the water at night.

The night was so dark one could not comprehend this: their huge bodies moving up that difficult slope. It somehow lodged in my mind that the darkness itself took their weight away, made them into ghosts. I felt like a ghost when I moved in the dark.

I took my meals up at the lodge. Every night I had the same thing, a plate of french fries with boiled cabbage on the side, served by a sweet-natured old Ugandan. He wore a tattered wine-colored jacket. There was still a skeleton staff in the park, different kinds of people tucked away here and there. And there were the remnant trappings of colonial Africa: tea was served in a squarish silver pot stamped with an elephant and a palm tree. I liked being there. I had been living on the road for a while and I liked routine. I liked eating the same food every night, its greasy blandness laced with vinegar.

Every night after dinner I walked in the dark back downhill toward my shack. I became aware that I was walking through a dense herd of hippopotami feeding in the grass on either side of the narrow dirt track. I became used to it. I was used to walking quietly in the dark down the faint-glowing stripe of the road.

One night I must have made an unaccustomed noise, scraped a loose stone with my foot. I heard a thundering nearby, somewhere down below, and knew I had dislodged one of the great beasts from that solid wall of darkness. I ran and ran and ran until it was quiet again, until the ground was still beneath my feet. And then I turned around and walked back down the hill to my shack, where I had left my things, my blue sleeping bag spread out on the bare bones of the stripped cot.

WHEN I CAME BACK to New York, Commager was dying. I went to see him in Sloan-Kettering. I was afraid to go. Lung cancer had spread through his body to his brain. I did not think I could bear to see him there, to see the disintegration of that dear face. But I went. I wrote someone a letter about it:

> I've just been with Commager in the hospital. There was a soft yellowish dent carved in the right side of his head, where part of his brain had been removed. Screwing up his one remaining eye, he said as I walked in, slowly, in surprise, "Miss Brind . . . I haven't seen you," then wonderingly, ". . . in years."
>
> There were pink lines sketched on his face. His fine bones, and paleness, and long brittle nose were beautiful

beneath them. "Yes," I said, "I just heard about this yesterday, or I would have come sooner."

"Just as well," he said, "I'm not a very edifying sight."

He spoke in sighs as always. "What! You've never read . . . ?" words stretched out and swallowed, as spasms traveled his frail, thin frame. I watched memory work in his face. His glasses slid sideways down over his bandages. . . .

Commager gave me a xeroxed page I carried with me for a long time. It was a copy of the two poems that end the first book of Horace's odes. Both of the poems were about Egypt. "Nunc est bibendum," the first began, "Now is the time to get drunk."

"Do not linger where the last rose delays," Horace advises in the second. What he really means to say, Commager said, is, This will break your heart, the richness of this beauty of Egypt with its atmosphere of great age and death.

What he really means to say is, Now is the time to get drunk. Cleopatra is dead. Died to escape being put on display, reduced to an entertainment for a cheap Roman pageant.

When Horace says, "I hate exotic things, woven crowns of linden," he is making a sad reference to Chiron the centaur, the son of the linden tree, who is described as laughing greenly as he comes out of his sacred cave in Pindar's *Pythian* 9. What he means to say is the holy centaurs of the hills are vanished.

I copied out a line from Pliny's *Natural History* on the page: *During the reign of Claudius we saw a centaur which he had brought from Egypt, preserved in honey.*

I had little interest in Egypt. But I began to study hieroglyphs in my second year of college. This was probably because learning them involved hours and hours of drawing. First you had to copy the text, reversing the characters from their original position running right to left. Then you looked up every one and made a list of what in each text the new words and characters were. There was a consonantal alphabet, and at the end of each spelled-out word a picture to suggest what it meant. All the characters were astonishingly concrete and in every aspect drawable.

I loved to draw. I had learned to draw birds by copying them out of the books of Fen Lansdowne, the Canadian naturalist. Aunt Dorothy had introduced me to his books when I was growing up. The Egyptian alphabet similarly contained a catalogue of birds. It was a naturalist's delight.

The letter m, for instance, was not simply an owl, but an eared owl, an eagle owl. The letter a, or alif, was the white Egyptian vulture, as opposed to the giant Nubian, or lappet-faced, vulture in the word for mother.

The primary consonant and symbol in father was the feared cerastes of the eastern Sahara, the horned viper, or, as we called it then, the male determinative, being also the designation for he, him, and his. You could see the little animal readily rising up sideways with its two horns into its sound, the letter f.

And there was the joy in discovering word cores, the threads that run throughout language, as in mut, both mother and death, and dwa, dawn.

Dwa is a picture of a star, a burst of light. Beside it in one common configuration is a human figure with hands raised in prayer. This was often translated into English as "Thank God." But it could not really be translated, I thought. It was too simple. The picture of a star, signifying this same word, was stamped all over temple and tomb ceilings as though it were a mantra.

The starry sky was a holy thing, a geographical concept, a country with a distinct topography of rivers and monsters and marshland that the soul would travel through at death. Where there were, I read one day, the lakes of a thousand geese:

"Wild geese come to thee by thousands, settling down in thy path."

In the old Egyptian texts a thing could be described in many different ways at once, and all manifestations were equally true. The king is cleansed in the field of rushes (the eastern stars at dawn). The king climbs to the sky on a ladder. The king becomes a gold falcon with emerald wings whose heart is from the Eastern Desert.

Sung to the great goddess Night:

one such spell began,

"Sew green stones, turquoise, malachite stars,
and grow green, that Tety grow green,
green as a living reed"

Thus, in the slow, painstaking process of translation, I absorbed these strange configurations without ever discussing them:

plant=life=light=star
death=night=desert=rebirth

I sat at the end of our dock in Geneva that summer, learning the stars, their Greek and Arabic names, searching out the rare green ones with my binoculars: Yildun, Sadachbia, Halcyone. I knew that rare green.

David caught a luna moth, or had my mother found it for him early one morning, pressed against the screen of the kitchen door at the cottage, a leftover from the night. Its huge, delicate wings were the most deliciously cool shade of green I had ever seen. Such a fine, unusual shade of green that I could taste it in my mouth, feel its delicate tissue on the tips of my fingers. How could I draw such a rare color into myself? I could only do this, taste and feel the color, as I looked at the moth.

My mother would never have let me touch it. She knocked it gently with the edge of something, a piece of paper, into her killing jar. It is on the wall of my childhood room in Geneva to this day, yellowish and crumbling in its frame of packed cotton.

I had a job as an intern in the Brooklyn Museum. I used to spend the afternoon in the vault. Its door was hidden behind a case of late dynastic sculpture in the Egyptian gallery. I sat sorting through glass frames of mounted papyri that had been lying around in dusty drawers for years. I sorted them into piles: Demotic, Coptic, Hieratic, all the written languages to come out of Egypt. Once in a while there would be a shred of something that I was unused to, like Aramaic, in the mix.

I liked looking at these frames of brown material that was more like fabric than paper. I had to be careful. If I held one the wrong way, a word could crumble and disappear. I did not understand the words. That was up to people who had worked intensively on these writing fragments for years, like the German papyrologist the museum would ultimately send the photographs of the sorted fragments. I knew what a difficult thing translation was.

East Wall · Sarcophagus Chamber
Pyramid of Teti · Saqqara

Words sung to the great goddess who strides across the sky.

"See green stones,
Turquoise, malachite stars.
and grow green. That Teti grow green

green as a living reed

That souls are made
of light · are stars,
are jewels are seeds
(expansive) are plants
and grow

Awe - light, plant, soul

نجم
زرع
(Bedouin) نور · See Toler me this in the
Hausa word 'to blossom'

The etymology of hieroglyphs is
visual as well as aural

· so, the papyrus
is green + the snake is green.
There are multiple levels of
meaning. The papyrus is
green because it is alive ·
thriving. the snake because it sheds its skin + travels it is
renewed in life, symbol and renewal. In the magic spell · in the
imperative "grow green!" · the papyrus in the world green has
sprouted new leaves.

ΕΠΟΣ ΠΤΕΡΟΣ · THE WINGED WORD

I was eighteen, working in the afternoons. The City of New York paid my subway fare. I was a VISTA volunteer.

In June of 1980, I went back to Geneva. I had been living in New York for six years, and had drifted into graduate school in classics. I longed for physical things, for the sun. I went to work in the peony fields beyond the Preemption Road. I showed up every morning at seven and lined up with the others to ask for work. At first I was always turned away. They couldn't place me: I wasn't a migrant worker, and yet I didn't seem to be local either, the kind of person they usually took on. Still I walked out to the fields every day. It felt so good to walk, freely (as one cannot do in a city), and alone, and in the early morning. I walked through the quiet streets and into the empty spaces beyond the town and the orchards of the Agricultural Experiment Station. In the second week they began to sign me up.

We went along the rows, bent over, cutting low on the long stems with jackknives, sprayed stickily head to foot with Off! Mosquitoes rose from the damp ground in clouds around our eyes. We shouldered the heavy stalks, their sharp-sided stems scraping through our shirts, and heaped them on tables to be sorted for the refrigerator trucks that would take them to Philadelphia or New York. On the breaks we lay flat on the ground where we were and did not talk. We smoked and stared at the sky.

There were always cast-off flowers. We could only cut those that were just beginning to go soft. The blossoming ones were too far gone, and the veteran cutters simply snapped their heads off as they went along. I walked away from the fields at the end of the day with a few dozen peonies. This was my real pay, for the money was little. I brought the flowers home, or over to Flo Whitwell's house on Castle Street.

Flo was one of the richest people in town, and, some said, fabulously ugly, with a cigarette usually in her left hand, a martini in her right, and beneath it on her thick pale wrist a very fine gold bracelet with the word bitch spelled out in diamonds. A present from her husband, Bill.

When I was fourteen, shortly after Barbara's death, Flo gave me one of her favorite books, an old Faber paperback edition of Edouard De Pomiane's *Cooking in Ten Minutes* (woodcuts after Toulouse-Lautrec).

With this book she hoped to state one of her own essential principles: that in life it is important to know how to do something with wit and flourish and style, for someone else, and to do it fast—like pulling a slightly nutty but wonderful meal together and having it on the table in ten minutes. It had to be good, which meant difficult, but it had to seem effortless. Most of all it had to be fun, because with that person you may only have ten minutes.

A sample lunch menu ran:

Skate with black butter
Beans *à la crème*
Potato salad
Cheese
Chestnut cream

For dinner:

Semolina soup
Calf's head *tortue*
Salad
Omelette *flambée*

"First of all I must tell you that this is a lovely book," Pomiane begins, "because I have only got as far as the first page. I have just sat down to write. I am happy. . . . My fountain pen is full of ink; I have fresh sheets of paper before me. I love my book because I am writing it for you. . . . The moment you come into the kitchen light the gas. Ten-minute cookery is impossible without gas."

That was Flo. She knew about cooking in ten minutes. She knew how to give a gift. Every time I went for the afternoon up to her house, where the back door was always open and one was expected to just walk in, I came away with some wonderful present: a silver salt spoon that curled into a ram's head, a champagne glass the stem of which was an opaque, tousle-headed Dionysus.

Bill would say to me as I walked in, raising his white head, his lean pleasant face (a face formed by wit, which is what we called him) to examine me through the lozenges of his gold-rimmed glasses: "You look like an unmade bed. What do you call this, twentieth-century attic?" (I wore, no doubt, a threadbare kick-pleat skirt, or something else my mother had made for herself in the early fifties.)

When Flo came back with my parents for a drink after a dinner at Belhurst Castle, she always brought me some wonderful treat: frog's legs wrapped in tinfoil, gooey lengths of sweetbreads cooked in wine. She sat on our living room floor and brushed out my long tangled hair. "If she would only get that hair out of her face . . . she might actually be . . . a pretty girl."

In 1980, I brought her avalanche peonies from the peony fields. She took me out to her garden and cut for me seven Siberian irises, my favorite flowers, she knew. I was so happy as I walked home down the long dip of Brook Street that I ate one. There were welts on my tongue for hours.

That summer I went fishing with David and his new friend the trapper Gary Lynch. Gary had trap lines running all along East

Lake Road. Like David, Gary was a naturalist: he knew where an animal was, how it lived, and how to catch it. He was David's first real competition in our part of the lake. Before they met they watched each other closely from their separate boats to see who this other person was who seemed to be catching so many fish, and, more important, where the pockets of bass were along the weed bed that season.

The weeds in those days came right to the surface, thirty feet up, and made a pale lavender flow of color beneath the water from a far distance, a lavender stripe running down the lake at that depth on either side. We knew of people who had drowned in the weeds, their lungs found stuffed with weeds at the autopsy, and we had always been wary as children of water-skiing over the weed bed on the final swing in toward our dock. That June, I spotted a fly rod David had lost standing straight up out of the water there while I was sitting on the dock after the sun went down, a faint black vertical in the cloud of yellow light.

That was the summer I first went to Egypt. At the Brooklyn Museum, I had heard about a man named Alan Moss who was just then starting a project in the Western Desert, a topographical survey over an eighty-mile area at the Dakhleh Oasis. My boss at the museum, a very sweet man who saw how badly I wanted to continue with the study of antiquities (and knew how difficult it was to make this a career), recommended me for the project. It was a gesture of great kindness, and he may have thought it would be just the thing to temper my budding enthusiasm for Egyptology. He himself hated working in the field.

I had no background in archaeology and no experience in the desert. Alan asked me to come to Toronto to talk to him about the possibility of my going. I took the train from the Finger Lakes around the eastern edge of Lake Ontario, changing to the older, better Canadian train at Niagara Falls. I stepped down to the platform in Toronto in the dark, in the pouring rain, and saw Alan waiting there, a short stout man with bright eyes and a black beard beneath a dripping, green-stemmed Japanese paper parasol.

"I hate cities," he said, as we rode back to his house in a cable car over the brick streets. Alan had worked twenty years before as an archaeologist in Nubia, cataloguing the fortresses and temples that were rapidly disappearing beneath Lake Nasser. Every year since that time he had spent a season in the desert in Egypt or Sudan.

"You must have a season in the desert." He laughed, as the car rattled roughly along. "It keeps you sane."

Alan had a grey clapboard house on Lake Ontario, a house in a row of similar houses on a back street. Tall red oaks with black-striped tight grey trunks stood between the back porches and the patios of the houses and the soft metallic blue of the lake. I was surprised to find in the city the Great Lakes shoreline of water-rounded stones, the color leached out of them so they blend with the water, and their silver rim of dead sawbellies among the washed-up seaweed (the bones pricking your bare feet, and everywhere the lake smell, the smell of fish slime, and dank weeds in the sun and gasoline). I mentally clung to that familiar shore as Alan stood in the kitchen cooking our supper and talked about the Western Desert, about the burnished black tracts of stone called "the melon fields," and the dune streets of the Great Sand Sea, and the barqans, the marching hills of blown sand.

I was frightened by the prospect of being in such a place, such an open place, and even more so of being there in a group of strangers. I would not be able to find there what I thought of as the one essential thing: privacy, the ability to go inside and close the door. I did not expect to hear from Alan again.

But he called me when I was back in Geneva and asked me to send him my passport. He said he would meet me in Cairo in August at the Garden City House.

SOLDIERS IN BLACK, lean and dirty, holding unsheathed bayonets, stood throughout the Cairo airport. The corridors leading in from the plane were lined with high iron bars. Behind the bars a mass of faces were crushed and screaming. These were the waiting families of the passengers. Only a handful of Westerners had come in on the Athens–Cairo flight that night, and I sought them out and hovered near them.

A young Egyptian plucked me from the crowd and rushed me through customs, out into the hot night air with its scratchy desert smell. The taxi, its radio screaming the popular radio lament music of Egypt all the way, dropped me at the Garden City House, an Italian pension on the third floor of a crumbling concrete building in downtown Cairo above the Tahrir Bridge.

In the morning I found Alan at a table in the dining room. All about him on the plastic tacked over the table was an assortment of canned jams, flatbread, boxes of white pasteurized cheese, and hard-boiled eggs. Rifling hurriedly, disinterestedly, through them were the dozen or so people who had just come from Canada or England to join the project.

Most of these people had been in Egypt before and had a humorous contempt for what the country had become: a film of filth over the monuments and sites they were interested in. I could understand their point of view. The city was so dense I couldn't see a way through it. I stood on the curb at the Midan Tahrir outside the hotel, my head throbbing in the hot liquid haze of exhaust, waiting for a break in the surge of traffic so I could cross the street. The Egyptians around me moved through the cars, picking up at once a sense of the rhythm of each car and almost dancing around them. I walked into the street, took a few steps, and looked up. A red bus stuffed with people, with people hanging from the doors and windows, was lurching rapidly, unevenly toward me. I froze and screamed. An old man jumped out from the curb, grabbed me and pushed me across to the other side, his face sweating fiercely as he smiled and laughed.

Alan loved Egypt, and he could waltz through Cairo. He had been born a clubfoot, someone later told me, and I thought, That's right, he knows something about the awkwardness of human suffering, that humor is the only way through it. In Cairo a cripple creates a traffic jam just by crawling slowly into the street. Alan used to say that what an archaeologist chiefly needs is humility.

I came down with a fever on the second day. I lay for a week in the Garden City House. I lay absolutely still in the heat, with the windows closed lest I be covered with flies. The flies had an unerring instinct for the dampness of eyes and skin. Outside, the traffic made a wall of noise, an impenetrable throbbing, like the sound of the ocean. Silverfish whirled lightly over the water-damp walls of the room.

A week later I sat in the back of a Peugeot station wagon driving west into the desert from Assiut. The fever had worked its way through my body, leaving me with an odd sense of stillness. I wanted just to look out the window and feel and smell the steady dry wind. We stopped at a sharp edge of rock where the land surface dropped abruptly away. This was the shore of an ancient sea. We wound slowly down to its floor far below. Sand swept over the road and, here and there, swallowed it completely. In the months that followed I would find on this sea floor that had become a desert pearly-blue oyster shells turned to stone, in heaps as high as hills.

My work was to walk and look. We were looking for remnants of whatever had once been alive in this desert, reduced to a polished trace of what it had been: an edge of an old mud wall, flakes of bones that had risen to the air after three or four thousand years and made a shining scatter on the surface of the sand like fresh snow.

One day I would find on a dull grey plain random stones that were the creased and twisting branches of trees, stained red with sap like blood. On another, crude axeheads that were among the earliest human tools, and flint bird points from the Neolithic that looked like delicately serrated leaves.

A limestone escarpment ran along the northern edge of the oasis. Its smooth white surface absorbed the penetrating wave of color that changed from one hour to the next in a sweep of intense disintegrating light. In the evening the sand seemed to melt in this light into deep pools of red and gold.

We were on the edge of the village of Al Aina, "the eye," the spring. The spring was an artesian well where yellow water bubbled up ringed with thick deposits of sulphur sludge. We swam here at the end of the day, sometimes in the dark under the masses of stars, and then went back to camp for a supper of roast turkey or goat, or to our beds to collapse in sleep.

The voice of a man walking through the village woke me in the hour before dawn. I could hear the creak of his swinging lantern as he sang out, with indescribable sweetness, the first call to prayer. I lit a match and carefully arranged my tent by lantern light: cot, blankets, mosquito netting, a reed mat, books on a shelf of split palm.

Outside, the sky was greenish-blue with the bright flake of Venus low on the horizon. I walked out into the cool, sweet air. I walked until the sun was directly overhead and unbearable, and, water gone, I sought in desperation for a rim of shade. I thought longingly on those first afternoons of my tent. When I got back to it I found it had collapsed in the wind, or, if it still stood, the bed, the books, what clothes I had were filled with dust and sand.

It took a long time to learn, and to learn bodily, that refuge was at best temporary: a shrinking line of shade, the coolness of water or the dark, in transient things like song or conversation. The vibration of a human voice is itself a soothing thing, and talk was like a blanket at night when we sat shoulder to shoulder for warmth, and I would walk into my tent and warm it simply by breathing.

Somewhere in the process of learning this, in the heat and exhaustion and the harsh openness of sand and rock and wind, my resistance gave way. I no longer looked for protection from these things. I walked barefoot in the sand and dove into a dune, feeling its warmth all around me. Bright-colored beetles popped out like jewels, amethyst, carnelian, and gold.

—

I spent most of the day walking with a canteen and a notebook toward the escarpment, which was always unreachable, in the far distance. A team of five or six local men came with me, carrying their hoes over their shoulders in case we found something to uncover. I would soon drop a distance away from them and walk off by myself. I knew I could always find them again by climbing a gebel, a small flat hill. However far away they were, I would see the bright envelope of their gowns and turbans in the shimmering atmosphere of heat.

My water would be nearly gone when I reached them a few hours later. When the men saw me coming they would stop and make our morning tea in whatever shade there was, often beneath a solitary tree. This tree would invariably be a variety of acacia, a gnarled, thick-branched thorn tree which also gave them twigs for a fire. The dissolving yellow flowers of this tree leave the pungent sweet smell in the air that is the smell of southern Egypt. A picture of an acacia pod is the hieroglyph for sweet.

The shade of the thin acacia leaves would transform the sky and its unbearable brightness into a lacework of delicate light. There was the heavy smell of boiling tea, and, if there was water anywhere near, scarlet dragonflies in the isolated pocket of green. We would sit for half an hour passing a small glass thick with sugar and tea leaves. I would strain to understand what the men were saying, and understood nothing. And then, without effort, began dimly to understand their words, and to speak, so abruptly that it seemed a mystery.

The first Arabic I learned was their daily singsong cry:

Taalu regala
niftur tahta shagara

Come, men! Let's have breakfast under the tree!

The escarpment had been our one point of orientation, and late in the season we climbed it. We found its huge truncated columns of limestone inset with whorls of white dissolving shells, coral, and fish bones. We looked below and saw the patterns the wind had made across the land, the dead rivers and the rows of squat, flat-topped gebels—small plateaus, that were heaps of hardened sea sediment, horizontally striped red and green with layered clays. Pigeons wheeled out together from fissures in the cliffs, dropped hundreds of feet, turned suddenly and disappeared.

In the evening we walked on liquid pools of light around the gebels. With pickaxes we dug out the sharks' teeth and fish bones encrusted in their sides. One evening we found the cracked carapace of a giant sea turtle from the Cretaceous, and once, the spinal column of a plesiosaur curled within a rim of flaking green mud. We followed it around the gebel, vertebra by vertebra, until we reached the cluster of its long yellow teeth.

At Dakhleh, I kept a child's arm on my desk. It was the color of amber. The skin curled like paper around the fragile bones. The fragility, the dryness of the object—were I to rub it, it would crumble—made even more remarkable the fine dusting of hair that ran over the skin. The fingernails, finely ridged like yellowing mother-of-pearl, pressed the evenly lined palm of the tiny hand, which closed inward, as though clutching something. Where the arm was severed above the elbow, the frayed tendons poked out like threads of white silk.

The child was buried in a Roman house. It was unusual to find a body within a house, and we made special note of the grave. The round buttocks, a shell of skin, the skull, where the flesh and hair had dried together and were flaking away, were just under the sand that had filled the foundation, a mud-brick grillwork now half a foot high. I picked up the arm, the one jarringly perfect remnant of the body, and wrapped it in my sample sack and took it back to camp.

It was a grace note, a trace of some lovely polished thing lying on the ground. The outstretched arm that held—was it a piece of bread?—in its curled hand. The hieroglyph to give.

Beside it on my desk a horned viper floated ribbonlike in a jar of formaldehyde. The draysha, as I later learned they were called in the Nile Valley (people there saying also that they will leap from the ground to sink their teeth in your throat and you will die in minutes, and sometimes that they are white and blind). A beautiful creature with a perfect soft set of tiny horns. The letter f, fy, the sound a hissing viper makes before it strikes. Someone had (luckily) found the little viper while it was asleep, beneath a broken pot.

One day Alan gave me a pomegranate. I had long had a fear of this fruit. As a child I had seen a black-and-white movie on television with my sister, a story about a cyclops, who is blinded in the end, a sharp stake gushing into his eye. Afterward we went into the kitchen and my sister ate a pomegranate. I took away an odd composite sense of the bleeding single eye of the cyclops and the ruby seeds of the fruit like blood in the tissuey yellow seed bed running with red juice.

At Dakhleh pomegranate trees were kept like great treasures in locked gardens behind high mudbrick walls. I knew the pomegranate had been cultivated in Egypt since ancient times. I knew this from the New Kingdom poem "The Dialogue of Trees." I had always loved "The Dialogue of Trees" because of the little sycamore:

> The little sycamore she planted with her hands begins to speak
> The flowers on its twigs are like honey
> It is beautiful
> Its branches shine greener than grass
> Its figs are redder than carnelian
> Its leaves are like turquoise
> Its bark like glass
> Its wood is green feldspar
> Its sap is opium
> It can lure anyone
> into its shade

But the poem began with the pomegranate:

> The pomegranate says,
> "My seeds are like her teeth,
> and her breasts are like my fruit—
> and I am the best tree in the garden
> I am always green . . ."

I ate the pomegranate Alan gave me, and then swam in the hot sulphur spring, delighting in the beauty of the experience:

the lavender sky through the palms, the bubbling hot water washing over my tired limbs, the sharp, peculiar smell of sulphur mixed with the faint smell of the decomposing reeds that surrounded the spring, the suddenly cool air of evening, and the color of the pomegranate, its fine brilliant red with its clear, slightly bitter taste.

When I came out of the spring my teeth and lips and mouth were black. I did not know this until the daylight of the next morning. Alan told me, laughing, yes, pomegranate juice and sulphur make a good black dye, don't they? I put away the little plastic-backed mirror from Cairo that hung from a string in my tent. In the end I didn't notice when the blackness wore away.

IT WAS WINTER when I came home. I arrived in Geneva on Christmas Eve. When I woke in the morning in my childhood bed, I didn't know where I was. Through the white curtains I could see the first red light of the sun staining the snow that lay under the maple trees and thick on their black branches. I had the impulse to get up at once and go outside, as I had done every day for the last several months.

I walked downhill to the lake to watch the sunrise. Thick smoky columns steamed off the shining crust of ice on the water. I felt a great shudder of joy. I could see this long-familiar sight, its penetrating beauty, with intense clarity. When I turned to walk back to the house there was a sharp pain in my feet. I could barely move them. I made my way slowly, determinedly, up the hill, up the icy sidewalk. On the thermometer pressed against the kitchen window I saw that it was fifteen below zero outside.

Christmas proceeded in our ritual of present-opening. After breakfast we sat around the living room, my mother and father, my two younger brothers, David and Charlie, and I, around a tree my father had cut at our woodland down the lake in Hector. David gave me a package that was brightly wrapped and very cold. I unwrapped it. It was a small frozen bird, a partridge. David was a very good wing shot, but it took him several Sundays in Hector to bring down the clever bird. He wanted, he said, to bring me a partridge for Christmas. I gave him a cotton-filled sample box with fossil shark teeth arranged in rows, the sharks' teeth I had dug out of the sea-clay gebels in the Western Desert. We used to find teeth like these on the shore of Seneca Lake when we were children.

The first days at home were filled with luxuries: hot water, lightswitches, mirrors, clean clothes. But gradually they became ordinary, and I began to feel that it would be a long time before I got away from them again, from their soporific comfort. I did not know how to find my way back to the Western Desert. I went back to New York, back to school.

In late February, David was driving down to visit me in New York. He stopped for the night at a college friend's house in

Connecticut, near the seashore. They went out for a beer, and his friend, while driving them home, crashed into a wall.

My father called me at five a.m. and in his broken voice said, matter-of-factly, "David is dead. I don't know what to do."

My mother asked me recently to show her the stone walls of Connecticut. I now live in the hill country of the Hudson Valley, some miles from the Connecticut state border. I told her I had never seen them.

There was a nightmarish struggle to get David's body back from Connecticut after the accident. It took days. There was some problem involving Connecticut state laws. Afterward I knew I could never live there. I went over the state line only to visit, and rarely as far as the sea, where David was killed.

David had stopped for the night to celebrate his friend's twenty-first birthday. In the weeks after the accident, the friend called my father on the phone, sobbing and incoherent. My father patiently listened to him and tried to talk him down, my father's instinct for kindness, and his lawyerly habit of listening, fighting (I thought at the time) with his impulse to kill the boy. Because his heart was completely broken.

My brother Charlie moved back into my parents' house from his fraternity at Hobart College up on South Main Street. He was a swimmer in those days, and I remember one loud pre-swim chant he and his buddies had before a meet: "It just doesn't matter." They shouted it out in monotone, over and over. I remember how good it felt to hear it.

When someone you love dies, you lose not only that person, but the peculiar combination of qualities they brought to the world. David was a naturalist in the old tradition: a hunter and fisherman. He knew how to track something, and he had the reverence and patience that come with this skill. I doubted I would ever know anyone like him again. Above my desk is a framed snapshot he sent me while I was at Dakhleh. Blond, thin, in blue jeans, one white-sneakered foot on the silver spine of a tuna three times as big as he is that he had hooked off the coast of Maine that fall. Nailed below the picture is the box of sharks' teeth I brought back for him at Christmas. These two small things have followed me from place to place since he died.

I woke every morning that early spring, and within seconds, as though someone had given me a good hard slap, remembered what had happened. David was dead. I would panic, and then settle into a kind of numb stillness in which I felt utterly alone.

In the beginning of the summer I packed up and left the country, having no idea where I was going. I wrote in my Shanghai journal, on a train going north through Norway in August,

Something like the Fall in me
All my leaves were dying
They died in the most violent way
And turned screaming colors

I was driven by a sense of instability that was inevitably helpful. There was no solid ground. I would have to go forward, stay on the move. I made the world a flood of running color through the windows of trains.

I traveled north to the Lofoten Islands in the Arctic Ocean. Coming south again I shared a compartment with an old Englishman who lived in Oman. He was murmuring out his lists of Arabic verb forms as we sped through the Dolomites. Listening to him, I knew that throughout the trip, curled on my sleeping bag spread out in train compartments, loving the rain and the cold, I had been heading back to Egypt.

It was not the idea of Egypt, or of Egyptology this time, that drew me there, but Egypt itself. Egypt as raw environment, difficult though it was. I have thought since that the difficult part of it was like a knife cutting away the frozen parts of myself. At Venice, I took a boat to Alexandria. I came into Cairo having shed my baggage all along the way.

ONE DAY IN CAIRO that year I went to the camel market in Imbaba where huge branded camels that had come from marching for months across the Sahara were being sold for meat. I watched as one, a scarred, furious male, blew a translucent bag of red veiny tissue out of his mouth like bubble gum. The men arguing beside him were immaculate, in a very fine layered white cotton. They carried daggers sheathed in snakeskin, and black wood canes. Their faces had a quality of wildness about them that I had never seen before. They eyed me with no reaction, scalding glasses of green anise or scarlet hibiscus tea pinched motionless between the thumb and middle finger.

They were tribesmen from the deserts of Sudan. I heard that they sold their stock in Egypt and traveled south again on the Wadi Halfa ferry across Lake Nasser. On their return trip they brought back from Egypt (as though it were the very edge of the civilized world where such a trade could be made) things they could not get in their own country: aluminum pots, battery-operated tape players and, most important, sacks of sugar, rice, and grain.

I wanted to know these people who seemed untouched by the modern world, who preserved their quietness, their integrity, even in the chaos of Egypt. I decided to go south myself. I got up one morning, left most of my clothes in a bag under the bed of the friend in whose house I was staying, and, taking very little in a knapsack, bought a third-class ticket to Aswan. At Aswan I went to the Wadi Halfa ferry.

I walked downhill to where the ferry was anchored below the High Dam. Behind a high mesh fence was a dense crowd of men, many of them traders like the ones I had seen in Cairo. I made my way among them and asked a tall young man if this was where we waited to board the boats.

"Yes, but you are a foreigner," he said. "There is no need for you to wait here with us."

Without another word, he lifted me up over his head and passed me to the man standing crushed in beside him, who handed me on to the next person. I was passed like a sack of grain over the heads of the crowd and dropped with dispatch on the far side of the fence. The army officers sitting at a folding table heaped with papers laughed as they saw me arrive thus at their customs outpost. They stamped my passport and said, "Have a nice trip." Such was my first encounter with the unshakable politeness of the Sudanese.

The ferry consisted of three battered rusted-out barges roped together with cables. I climbed up the wet wood plank and found myself alone on a floor of metal plating and splintery wood. A metal staircase led to an upper level, where I made a place for myself beside the railing. I knew I would be there for a few days, and wanted to be able to look out over the water. I set up a space the length and width of my sleeping bag—David's duck-hunting subzero sleeping bag. It was pleasantly soft and thick and served as a carpet as well as a bed. I wrapped all the clothes I had in a green-and-black-striped Negadan scarf I had bought in the Aswan suq the night before, and this became a pillow.

Everything I had with me—oranges and chocolate and a bottle of water; paper and a bottle of ink; two silk scarves, one white, one peacock blue, shredding at the edges and streaked with watermarks—I set out as an arrangement of color, as though building frail visual walls to close myself into a small familiar world.

The Sudanese on the boat, I soon discovered, far excelled me in this practice. When we left in the late afternoon, the floor was spread mosquelike with pastel rag cotton rugs. There was not an inch free of people. Yet they had settled naturally into geometric patterns, and I did not have, even for a moment, the sense of being trapped in a crowd. Men reclined in circles smoking honey-soaked tobacco in water pipes. Wands of incense burned throughout the open level. The railings were hung with the gowns the traders had changed out of. These made curtains of wet translucent white cotton filled with the comforting, delicious smell of laundry soap.

A calm washed over me that evening with the gently moving shadows and patches of light, a calm more powerful than a drug. I did not know where I was going. No one who knew me knew where I was. The other passengers on the ferry quietly accepted my being there and left me alone.

In the weeks before, when I said to my friends among the peasants in southern Egypt that I was going to Sudan, they would invariably respond,

"Are you going down to buy a slave?"

"You're going to Sudan? You will surely be eaten. The people themselves will eat you."

"The Sudanese are black. They are scarcely human. They don't even speak Arabic. They bark."

As I sat among the Sudanese tribesmen on the Wadi Halfa ferry I was struck not only by their innate sense of decorum, but by their physical beauty. They seemed to have absorbed elements from all the different races at once so that one saw in a single face the slanted eyes of the Chinese, the thin, delicate bone structure of northern Europe or India, eyes that were shades of lavender or mint green seen nowhere else on earth, and skin that ran from aslia (the color of honey) to copper red, to an ashy greyish black.

Sudan, elbelad asswadeeen, "the land of the blacks," was a market term, I suspected, a term of the Arab slave trade. It was a bad joke, for the territory itself was an argument against the notion of race. It spoke of nothing but the astonishing diversity, fluidity, and adaptability of human life. The name Sudan referred to a vast area that was marshland and jungle and river basin and desert and mountain and tropical sea, the Sudanese a hundred tribal groups, each with its own language and physical character. The Sudanese did not commonly use black to describe a person's skin but yellow or red (Mahas, Danagla), green (Kordofani), or blue (Dinka, Shilluk).

The mixing of blood had gone on in these parts since prehistory. The Red Sea coast was utterly porous. The Phoenician, Greek, Chinese, and Indian traders who regularly laid anchor in the treeless harbor towns with their shining white walls of stacked coral and their shrinking water supplies—towns that dried up and vanished and are now so many mysteries, so many archaeological sites—left not only their smashed pottery and coins among the stones of the shore but traces of their blood throughout the region. They left their histories in stories that are still around, about the Rum, a giant white race that once ruled the coast. About the Kingdom of Axum and how it had tried to storm the citadel of Sheba across the sea and failed, and gave the Quran the wonderful line, "Alam tara kaif faala raabaka bi ashab el fil?" Did you see what God did to the friends of the elephants? For elephants were brought from the inland forests and shipped out for military purposes along the Red Sea.

The Phoenicians came to the coast early on and, some say, gave their name to it. Phoenix is a Greek word for red. They may also have given their name to Punt, the paradise somewhere south on the Red Sea. Hatshepsut, the woman pharoah, sent an expedition south to find (after a thousand years of rumors about it) the exact location of the land of Punt, the source of the great luxuries of ancient Egypt that came from outside the Nile Valley, the incense and precious woods and corals and jewels. The story of the expedition is illustrated on a wall of her temple at Deir al Bahri. The sailors come to shore and find the king and queen of Punt, a red-skinned bearded Semite and an African woman ringed with folds of fat. "How did you find the way," they say in hieroglyphic fumetti above their heads, "to God's own country? Was it by water or the wind?" The sailors come away with ships filled with gold and ivory, and trees in pots. Monkeys hang along the rigging of the ships that sail up the Red Sea, its water thick with squid and rays and odd-colored fish etched in stone in gorgeous detail.

The location of Punt was still being argued about when I was in school. Was it the coast of Sudan, or the Yemen, or was it Oman, where the terraces of incense trees were? My learning piece in hieroglyphs was a Middle Kingdom folktale called "The Shipwrecked Sailor" that predates Hatshepsut by several hundred years. The story is about a sailor who is shipwrecked as he sails south to the copper mines on the Red Sea. He is washed ashore in the land of Punt, an island that seems to have "everything that exists in the world." A giant serpent threatens his life, then learns that they have suffered the same terrible fate, the loss of their companions and loved ones, and becomes his friend.

In the end the sailor sails away with gold and ivory and myrrh and apes and giraffes. The serpent says to him, "After you leave here you will never see this place again. It will become water." Such was the truth about the land of Punt, I thought. It did not exist. It was made of wind and water, a cover for the secret of the tradewinds that brought ships back and forth between India and Africa, carrying with them goods from all over the world. And carrying the Phoenicians, and whatever traders succeeded them in power until *The Periplus of the Erythrean Sea,* the sea log of an anonymous Greek sailor, revealed the secret of the spring and autumn winds that sweep east and west across the Indian Ocean. The story of a paradise somewhere south on the Red Sea continued on as a great riddle. No harsher desert country existed, and many people came to the Red Sea coast and died of thirst.

The inhabitants of the coast were a people who had been known by many different names. I had been reading about them for years as the icthiophagoi (the fish-eaters), the troglodytes (the cave-dwellers), the fuzzy wuzzies, the Beja.

"For the forty centuries of their known history," a British official wrote of them in the 1950s, "they have watched civilizations flourish and decay, and, themselves almost unchanging, have survived them all. . . . As a record of survival it is indeed unique, attributable partly to the inhospitality of the country in which they live, but more particularly to definite traits of character, preserved almost intact by their free, nomadic way of life."

Around me on the boat were descendants of the Beja: Hadendowa, Ababda, Besharin. They had come into Egypt to buy or sell, and left it casually, as though it were nothing, for their own roadless, difficult home country.

Evening fell. A boy swung up on the roof of the boat from the railing below and sang out a sweet thin call to prayer. A little while later an old man with curly grey hair and the chevron scars of the Mahas Nubian tribe looked up at me from the lower deck of the adjoining boat through his thick glasses.

"First time in Sudan?" he called out in English, laughing. "Will you be surprised. No food, high prices, lousy government. Wait there. I'm coming up to join you."

Dr. Hatikabi introduced himself as a poet and said that he had a bookstore in Khartoum. He was on his way home from an operation in Cairo. We found we knew some of the same people there.

He came up to sit with me and asked me what I intended to do, inasmuch as I seemed to be completely by myself. Where was I going? When he found that I knew nothing at all and had no plans, he drew me a map of Sudan. He marked it with all the tribes across the country, tracing out the patterns of their scars, which varied dramatically from place to place.

If you want to see the strangest-looking people on the earth, he said, you must go to the south, to the Sudd, the marshland of the Upper Nile. In the Sudd there were people with asterisks pricked over their whole bodies. There I would see things, Dr. Hatikabi assured me, that I had never seen before in my life. He began to teach me old Sudanese songs that he knew. The music sounded strange, when I first heard it, antique, more Indian than Arabic.

As he sang, tribesmen came and sat around us. I began to understand that year about trading poems and songs. It involved giving, that intangible, freeing human thing: giving something priceless, even to a stranger, for nothing.

A few months before, I had sung to a room of Egyptian engineers who were building an aluminum factory in Edfu. They had given me dinner and had sung to me, tapping their glasses with their forks for rhythm. Then they waited, expecting me to sing something back. I sang "Mean to Me," belting it out. They all laughed wildly and applauded on "honey," the only word in the song they understood.

A decade later I was with my friend Nina West in the Tien Shan Mountains, between Kazakhstan and Xinjiang. Everywhere we went we sang—on buses, in the high rich green mountain fields, walking along a road. And in response, everywhere we went people sang to us. They traded beautiful Kazakh and Uighur songs for "You Go to My Head" (Nina's favorite), and "If Tomorrow Wasn't Such a Long Time" (mine).

One night in the snow at Heaven Lake, in a concrete shack where we fed together on a sheep's head, we started to sing Beethoven's "Hymn to Joy." And to our surprise everyone else in the room sang with us—Russian, Chinese, Kazakh, Mongol—not the words, but the music, for everyone knew it.

On the Wadi Halfa ferry I began to write down what people said, what they gave me, or to seem to write it down, for it was usually way out of my depth. It was rarely the song or poem itself I was interested in. I wanted to know the person who was giving it to me. The giver and the beautiful gift. But somehow, writing down what they knew by heart, the writing itself, which was such an unusual, an unnatural thing, seemed to make the words significant in a new way, and prompted them to give me even more.

I would recite a verse from the eleventh-century epic about the Muslim conquest of North Africa, the *Sirat Beni Hilal,* that I had heard in southern Egypt that winter: "Shabahat nefsi kasfina saghaira . . ." "I compare myself to a small boat on a rough sea. The waves lift me and drop me, and the water is full of darkness, and its waves are bitter." Sometimes I would receive in response another verse that the listener remembered—and there would be the shared delight in recognition, like a sudden burst of light.

I bought a child's Quran in the Aswan suq the night before the ferry left. It was a collection of the short suras that everyone knew by heart. I had never read the Quran before. But I had often heard these simple verses and thought I would use the time on the boat to memorize them. I had been brought up to memorize poetry. My mother knew astonishing amounts of poetry from her childhood, long passages from Shakespeare and Tennyson (and would recite them at the slightest provocation while Barbara and I would squirm to get away). I could never match her memory, or Barbara's for that matter, but my head was full of words, poems, poetic lines (and often, like my mother, I couldn't quite remember where they came from or how they had gotten there).

When I lived on the road, I found I didn't miss familiar things or books. I couldn't carry very much with me. There was instead great pleasure in simply sitting and remembering. I thought of memory as a blanket. I could take a thing out of my mind and handle it as though it were part of some beautiful fabric I carried with me, things that had happened long ago, the faces of people I loved, the words of a poem I had long since forgotten I knew.

This was something any nomad or illiterate peasant knew: the intangible treasure of memory, of memorized words.

The Meccan suras were where one started. They were simple, clear, curious, and wonderful. They spoke of a refuge where there was none, where no physical refuge could be found:

"I take refuge in the lord of the dawn, from the evil of all created things, from the evil of the growing dark, from the evil of those who blow on knots, and from the evil of the jealous in their jealousy."

And they were powerful. One morning on the boat I stood looking out over the railing at the dawn. This was on a different boat from the one where I slept, the pilot boat, and I stood outside the engine room, where I could always get a glass of tea in the morning. I stood with a glass of sweet steaming tea looking out over the water, and the navigator, a Nubian, came with his glass and stood beside me. I had not spoken to him before and he said nothing. We were watching the sun rise.

As the edge of light appeared he said, "W shams w dahaha"—

By the sun as it shines—the first line of one of the Meccan suras. And I responded with the following line (for I had begun to memorize the sura the day before): "W qamara iza talaha"—By the moon as it follows her.

At Wadi Halfa the boat came to rest and, quickly, became the sad wreck it had been. The traders moved in a long, slow line to the train track a mile away, their huge white sacks pressing precariously on their hunched backs as they made slow progress over the desert ground in the hot sun.

There was no road from Wadi Halfa to anywhere. Only the train, and that went to Khartoum. Its wood slat seats and glass windows were broken. There were no lights. As we left in the evening, dust filtered in steadily over us. It gave a thick amber quality to the last of the sun. As the train moved on, I could not keep the dust from thickening on my lips and making its way into the dampness of my mouth and nose and eyes.

I understood for the first time the veil. It was the one thing soft enough to shield the face. There was no way to think you could not endure the suffocating coating of dust. This was a fundamental aspect of life in Sudan: the inescapable physical world with its discomfort, always the near presence of pain. And the need to rise above the pain, to expand into a safe place even when there is no space around you.

The car was crammed with people. People from the boat slept in heaps in the aisles, between the seats, on what remained of the baggage racks. Their bodies curled in white gowns, blurring with the whiteness of their sacks.

During the night the train stopped. I waited, straining in the silence. Hatikabi appeared. He made his way carefully back through the car. He said we had reached water station number six, and could climb down and have some supper.

I could see along the tracks in the blue light of the train. Banks of sand sloped down on either side to where men squatted around aluminum tubs full of beans boiling over small cookfires. With the hundreds of other people from the train we slid down into the sand.

We sat silently eating the warm food in that moment of stillness. Abruptly, the brakes lurched above us and the train began to move. Hatikabi grabbed my arm and ran. Everyone was running and leaping at the side of the moving train. I leapt and fell, and ran again, and caught an edge of metal. I held on and swung myself blindly into an open door.

I never saw Hatikabi again. Or the sack of clothes I had left on my seat. But I had learned the trick of riding on the train. I sat in the cool clean air of the doorway, where the wind washed the sand and dust away over me. I looked out at the familiar stars.

—

I took Hatikabi's advice and a month later, having gotten off the train at Abu Hamed and crossed the desert to Khartoum to continue south, found myself on another ferry. The long curved horns of a bull were roped to a mast above the pilot's cabin. We were at Kosti, waiting to travel through the Sudd. This was a longer trip than the first, roughly two weeks because it was March, before the rains, and the Nile was low.

There were perhaps a thousand people on the Kosti–Juba ferry when I took it. I read in the newspaper when it sank two months later (when it was attacked by bush guerrillas above Adok) that more than six hundred people drowned. There was no room at all. Bodies, and everything attendant upon them—food, human waste—were everywhere. Still, there were places on the boat, if you moved around and found them, where it was pleasant to be.

I learned to sleep on the roof, where it was cool, and I could see at dawn the strings of waterbirds in flight that seemed to draw up the sun. They drew it up in a golden swell that soon became scalding, white, and burned the color out of the day. It burned my skin so that it fell away in shreds.

During the day I climbed below, where people were gathered over their meals of grapefruit and onions, and yellow eels that hung strung up over the sides of the boat to dry. Thirst was constant in the heat. I learned to toss the powdered-milk can that lay on deck, fastened with wire to the railing, over the side; I pulled up and poured the cool, muddy Nile water over my head, soaking my clothes.

One day we crashed into a mango tree, and its full load of fruit came scattering down over us like a blessing. We scrambled to grab as many mangos as we could before the owner of the tree, a little man in rags, leapt onto the boat with his heavy stick and began to beat people back. But he had lost his crop.

The southerners around me had skin so dark that it seemed almost tinted blue in the intense sunlight. It was twisted and cut in the patterns of rows, knobs, and swirls that mark the Dinka, the Baqara, the Nuer, the Shilluk. All of us could use Arabic to some degree, each with our own dimly recognizable version of it. I would say to whoever sat near me on the roof, in the Sudanese dialect of Arabic I was hearing, "Tair da shinu?" What is the name of that bird?

Pteros, tyur, Altair, the name my mother taught me for the constellation Aquila. Altair, the sapphire star, in its torn and fading triangle in the upper sky, Aquila, the eagle. I could always see it, even as we neared the equator, and the bright lopsided Southern Cross.

There were always birds around us. The tall goliath heron and the fish eagle were our companions all the way. They flew—tense and slow on their giant wings—alongside the boat to snap up whatever our progress stirred to the surface of the water. There were small black-and-white birds with streaming snake-tongue tails three times the length of their bodies that spun circles behind them in flight, and insectlike bee-eaters the colors of precious stones.

The river broke into streams that wound through traveling fields of high papyrus, green and golden tufted in the sun. African water hyacinths, deep purple, soaking up light, and climbing vines of pale veiny flowers surrounded them. At night bright crowds of fireflies rose and fell through the dense screen of vegetation, doubled by their falling shadows, as we wound slowly on, stopping frequently because the depth was so low.

One night we hit a sandbar, and tiny birds, the kind that clean the mouths of crocodiles and hippopotami, covered us as we slept, in a panic plucking at our clothes.

We had entered the great papyrus marsh that is etched in stone in early Egyptian tombs, the primordial swamp that was Egypt. The swamp in the desert where life swells and proliferates, in all its wild variety, out of control, a swamp of paper. The thickness of the scented air was intoxicating. There was no sense of progress. Our path was obscured by varied sun-threaded greens that stretched to each horizon, swallowing the narrow lines of water every hundred yards.

After some days the boat made its way to Juba, where I left

it. From Juba I continued south on the back of a coffee truck driven by a Kenyan named Job. The other passenger was a young South African who had recently been a soldier. We held on to a rope to steady ourselves as we bounced violently along the red clay roads into Uganda, racing to beat the rains. The rains came, and we slept on the cold metal truck bed in the downpour. The ragtag Ugandan police arrested us in Gulu. Job bribed them to let us go. The truck broke down in the Ugandan bush west of the Kenya border. I found a jitney to take me into Kenya. At the border I heard that the rest of Job's convoy had been ambushed by rogue soldiers on the road. Months later, traveling north through Uganda, I saw the burned-out shells of their trucks.

I was heading back to Egypt. I longed to go back to Egypt. In the years that followed I went back to Egypt again and again.

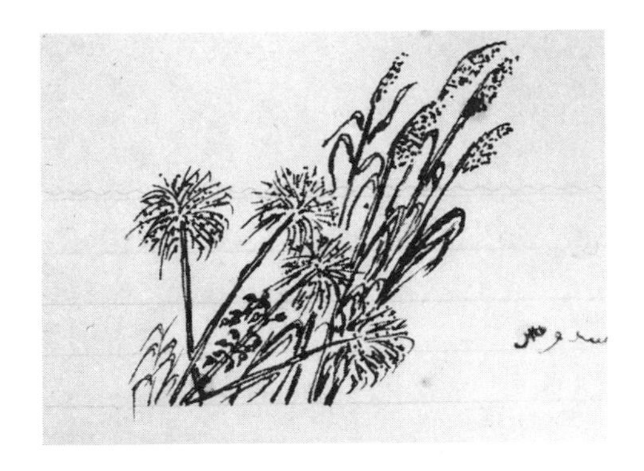

CAIRO | 1989

CAIRO IS UM A DUNYA, Mother of the World. Coming into the city I am appalled at how filthy and chaotic it is.

I arrive at the airport, which is in the high desert, and descend into the Nile Valley where the chaos of human life spreads like a dense fog through the streets. Peasants with donkey carts, women in shiny synthetic Islamic gowns, cripples on crutches and wheels, young men in Western clothes all wade through the moving cars without seeming to see them. At times in Cairo there is no longer the sense of individual people moving here and there, but of an undifferentiated mass of life: a single organism that has bred huge and covers this patch of land where the desert meets the river.

The rattlebox taxi weaves unevenly through the traffic of Heliopolis and into the heart of the city, crisscrossed with rusted-out raised walkways. It takes nearly an hour at midday to arrive at the Giza corniche, where I get out.

As I come down the narrow concrete stair to the river (beyond the heavy iron gate, painted blue. It creaks painfully as I shut it behind me), Galal leaps off the boat onto the grassy bank, face turned up toward the bright afternoon sun. He shuffles up to me, hands in pockets, taking the steep stairs lazily, grinning and whistling.

When we last saw each other, three months ago, he hugged me stiffly in tears as I climbed into a black Mercedes limousine on the corniche before dawn. The normally mobbed street was empty and still save for the first bird noises, sparrows clicking like typewriter keys in the sycamore leaves, the shriek of the invisible kerawan.

I was flying to Istanbul to meet my husband, whom Galal has never met. His only knowledge of my outside life is of the harsh metallic ring of the international calls that come through in the dead of night.

I have been living on this old steamer in the Nile with Galal, a tiny old Nubian, three scars knicked in beside each eye, on and off for a year now. We are anchored in the heart of Cairo, just south of the University Bridge.

Cairo is a river city. Naked boys jump off traffic bridges to swim in the brown water of the Blind Nile. Peasant families live on the banks, their belongings strung up in the ancient whitish fig trees. Families of fishermen live in brightly painted rowboats, singing and drumming up the fish into the nets they spread across the river through the day. When there is a good wind they use blankets and rags for sails.

Galal calls all the small fish, the fish they catch, bolty. They look like perch, but somewhat greyer. He fishes over the side, his legs propped up to steady the busted wood chair he sits on, rocking back on its three legs.

Galal does not like the fishermen. He throws my empty tonic water bottles at them as they row by us in the early morning, working the heavy square oars with their whole bodies. "Oh, Om Galal, Uncle Galal, morning of blessings, morning of light! How are you this morning, Uncle Galal?"

A bottle splinters on the wood hull with a crash or plops into the water beside them. "Utla bara min hina! Utla bara min hina! Utla baeed!" (Get out of here! Get away from us!) They splash him and laugh and come back a few hours later.

Though there are "good" fishermen, too. The walleyed Ibrahim who has a red boat with black trim. Sometimes I come down to find him watching a soccer match on the little black-and-white TV in Galal's cabin, denlike and dark beneath the stairs, a water pipe clicking between them. But generally, Galal says, "kulahaga fihum"—these people are capable of anything.

The boat was once a paddle steamer with Thomas Cook tours. In the thirties it carried small groups of wealthy English tourists up and down the Nile. There is only the dimmest suggestion now of its former life. The blackish fittings are brass. Uneven white paint covers what must have been fine dark wood in recessed paneling. The bathtubs are enormous.

But now the old wood flakes and bends, and the boat looks like all the other houseboats anchored along Giza and Imbaba, sinking deeper into the river every year. The houseboats are fabled for orgies and hashish and rats, but for the most part, like any free space in the city—tombs and rooftops and the undersides of bridges—they house squatters and the poor.

Fustat, as this one is called, passed from Thomas Cook into the hands of archaeologists working for UNESCO in Nubia before the Aswan Dam. Now the American Research Center in Egypt keeps it for visiting scholars and archaeologists. American researchers rarely get permission to work in Egypt these days, it seems (or at least those who have have completely lost interest in *Fustat*), and Galal and I usually have it all to ourselves. The Center hopes to sell it to a Saudi sheikh. "But let's not talk about it," Galal says. "That will surely make it happen."

Galal has a violent temper, a psychotic streak. His rages successfully frighten marauders away. He is a little man with a reedy, gravelly river voice. His clothes are oil-smeared rags. A black leather pilot's cap and thick square black-rimmed glasses hide his grizzled face. He is very funny. He talks a lot. He plays with words. Like most illiterate people in this country, he has a brilliant sense of language. He brings me odd words, wordplays, jokes, as gifts. His first language is Nubian, an African language rich with reduplicatives—but this is his language from long, long ago, before his country vanished beneath the dammed river.

The dammed Nile is a dying river. Here in Cairo we watch it daily pumped full of pollution. Blocks of Styrofoam drift by like melting snow amid animal corpses and submerged tangles of colored plastic. Galal pulls animals and birds out of the black oil that often laps up around us. He is always nursing some sick little creature, and now has three hand-sized kittens on the lower deck to whom he is feeding milk. He peers up through his thick glasses and hands me one. I can feel the ribs, and heart throbbing through the skin.

The upper deck is mine. We begin our usual routine of opening it up. It has been empty and closed since I left at the beginning of the summer. Galal goes into his cabin and brings out the keys and an axe. We open the door to the musty dark rooms. The hard part is opening the windows, working the axe into the cracks of the blue shutters, warped shut with the dampness and heat. Galal pries them open one by one. There are six windows, three on each side. I attempt the same with my Swiss Army knife and the blade snaps in half.

I roll up the dust-filled synthetic pink carpets and stuff them in a closet, and take down the familiar pictures of Egyptian temple statuary (that Galal always restores when I leave). Giant cockroaches leap out from behind the frames. We rush to squash them before they scurry under the beds. This leaves only the old faded maps on the walls, boundaries and names all wrong now, of the Middle East, Egypt, Sinai, Sudan.

I eccentrically insist on pushing the two metal cots together to make a broad double bed beneath a window facing out on the river. I shake out the sheets—stained with dead bedbugs—and put on clean blue cotton ones, unzip my blue down sleeping bag as a quilt. Beyond the bed is an old oak desk, by the cabin door, and on it a cedar box with last year's collection of desert rocks.

Rats have eaten the leather bindings off my books. "We will have to call Om Sulaiman"—Uncle Solomon, the bookbinder. Galal laughs. He goes off to forage above in the streets of Giza for the usual things: a newspaper funnel of oranges, another of arugula, lettuce and dill and little green lemons from the peasant women who sit on the sidewalks draped in black, their lips and chins tattooed blue.

He brings a dozen roses, small, tight, orange, like the roses of Alma-Tadema, and a small bag of chocolates from the Saudi Hotel. Paid for this effort, he goes downstairs to smoke and pray, and returns later in the evening with two glasses of stewed sugary tea.

—

The ritual of arriving in Egypt: excavating a space in the decay of an old magnificent thing.

I run a hot bath in the huge bathroom, where a back wall of warped mirrors reflects the river light through the frosted glass.

A water smell, fish, the smell of the sea. The boat rocks beneath me. The bathwater splashes onto the floor. I lie out on deck in a tattered old Ritz-Carlton bathrobe. When the steam stops rising from my skin, I go back to the bath again. I spend the day in and out of the bath, watching the city move all around me, the wild, impenetrable business of the city. The air is thick with a chemical haze. It veils the other side of the river.

A rush of wings as thirty white egrets fly over me and land on loose clumps of water hyacinths floating by.

The clearest liquid green flows into the sky. The first bats are out. A boy on a neighboring roof (he lives on the roof) is calling down the pigeons. First with a long ragged piece of red nylon, a flag, he waves and jumps, whistles, hissing. Four, finally a fifth are circling him. They will not come down. He takes a stick, square and blocky, waves it in the air, beats the roof, still whistling. At last they come down and are still, blurs of white in their box nests. The sky as it deepens into blue is the loveliest color in the world. The first star pops out into it. A kestrel dipping over the river babbles, wings whipping limp and taut like ribbons in the wind.

I CAME TO CAIRO in 1988 as a fellow of the Crane Foundation. An auspicious circumstance, I thought, for the crane was a migrating water bird, a bird that was associated with the origins of writing. Palamedes invented the alphabet after watching the patterns that flocks of migrating cranes made in flight against the sky.

The crane was a multilevel symbol. The flight of cranes marked the turning of the seasons, and marked the remnant sources of water in the desert where the birds came down to rest on their passage between Africa and Asia. Its name was an echoic root, an imitation of its cry: crane, crow, kurki, heron, kerawan, geranos.

As in the third book of the *Iliad*:

> The Trojans went forth with a cry and shout, like cranes,
> like the cry of cranes when they fly toward heaven
> fleeing winter and the winter rain.
> They fly with a cry along the streams of Ocean,
> bringing death and destruction to the race of pygmies.

The crane was an auspicious bird, and I was happy to come to Egypt under its protection. I had been coming to Egypt for years with no other desire than simply to be there, but the origins of myth, the roots of words, insistently presented themselves to me. Egypt was the world's great treasure-house of the manifestations of such things over time. It not only revealed in casual display the kaleidoscope of manifestations, but led one haphazardly in the direction of their source, the natural world.

I was, I thought, in my way, a tracker like my brother David. I was out to find the living thing in the wilderness. The wilderness was, after all, where things came from. In Egypt this was ever that frail edge where the desert meets the river or the spring or the rain-washed wadi bed, where Moses (in the Coptic etymology of the name) was the conjunction of two primal words, two essential elements: mu (the letter m, in its earlier form the hieroglyphic picture of a wave), which means water, and sha, which means tree; where stars and flowers were described with the same words, with the Coptic reduplicative shasha, the hieroglyph for a bed of waterlilies, which means to blossom, and to appear in the dark.

In this wilderness you might encounter the horned goddess whose horns were at once the waxing moon bringing water, and the horns of the wandering cattle the herders who once lived in the Egyptian desert followed in search of the seasonal pasture that water would bring. They were the horns of Hathor, the cow goddess of the desert mountains, whose holy ground was the Red Sea Hills with their caches of precious stones. You could imagine them forming under her wandering feet like flowers around her temples at the mining sites of Serabit el Kedim in Sinai (Sinai a name which means province of the moon, of the cow goddess, the golden cow), and Sikeit at the foot of the Emerald Mountain: the precious stones that were flowers mirroring the jewel-like colored stars in the sky through which the delicate moon wandered marking out the patterns of the year, the months that brought rain and drought. They were the horns of the desert demon, the Queen of Sheba, and the horns of the statue of Isis the Beja tribes, once cowherders themselves, carried away from the temple of Philae at Aswan when the newly Christianized Romans tried to close it down in the sixth century.

You might, in the Egyptian wilderness, if you were lucky, meet Mari Ghergis, St. George, also known as El Khider, "the green man," the endless possibility of the renewal of life, who wanders the desert and appears to those who are lost and in despair. Or you might instead find Lilith, the terror, the goddess of the parching wind.

The figure who stood between the wilderness and civilization was King Solomon, who was, in the folklore of the Middle East and in the Quran, a shaman, a magician. The wisdom of Solomon was the wisdom of Egypt: the ability to interpret the natural world, to translate nature into language—Solomon's magic mirror that revealed all the places in the world at once.

Solomon controlled the winds, and he knew the mantiq a tayr, the language of birds. The mantiq a tayr had fascinated me for years. There were many versions of the story, in the Quran, in Aristophanes, in the poem by the Sufi Faird uddin Attar (which is how I first knew it, from a beautiful seventeenth-century illustration of this poem from Herat that I cut out of a book when I was fourteen, a tiny painting on gold-edged paper, of birds beneath a sycamore tree on a lavender ground).

All the stories involved the hoopoe. The hoopoe is the distinctive bird of the Nile Valley, and is such a delightful creature that it could easily have given rise to many different stories (although the ones I knew went back, no doubt, to the same unwritten tradition). The hoopoe is one of the creatures on the earth that seems to have been made by someone with a terrific sense of humor. It is mottled, and at first glance looks like a giant moth, with its striped wings of brown and black and white spread in a slow, floating motion low over the ground. The feathers on its head rise hesitantly up into an Iroquois crown. Its face is sweet, and ends in a beak that is long and curved like a sword from *The Arabian Nights*.

In *The Birds* of Aristophanes, two men in debt leave Athens to found a new colony, Utopia. They first set off to find the hoopoe, who has flown high enough to have seen the whole world and will know where a good place will be. Ah, the hoopoe says to them, I know just the place you're looking for—somewhere south on the Red Sea.

The men persuade the birds, who are the first gods that ever were, to rebel against the domination of humankind and take over the world. The Greek speeches in the play imitate the sounds that birds make, their "language," and contained in the play is the idea that the language of birds is the earliest religion, augury.

In the Sufi poem "Mantiq a Tayr," the hoopoe persuades the other birds to fly off in search of paradise. After many trials and much discouragement they find it: a mirror reflecting themselves.

In the Quran, Solomon knows the mantiq a tayr. The bird he most relies upon is the hoopoe. This is because the hoopoe can see water under the ground, and so knows where water is throughout the desert. The hoopoe has flown south along the Red Sea and seen there a queen who is as beautiful as the morning star and has great wealth in gold and precious stones. The Queen of Sheba lifts her skirt to walk on Solomon's glass floor, mistaking it for water. In doing so she reveals her hairy legs, the legs of a wild ass, and so betrays her identity as a desert demon.

The Red Sea coast was the wilderness, where Solomon exiled the demons who had built for him the temple in Jerusalem. The Beja were descended from such, from a jinn named Hafhaf, hot wind, or Sakhr, hawk, "who deceived King Solomon in the matter of one of his wives."

In 1988 I came to Egypt on a two-year grant from the Crane Foundation with the intention of going to the Red Sea Hills, to the illusory paradise somewhere south on the Red Sea, to the stretch of mountainous coastline between Egypt and Sudan, the Etbai, which was the tribal territory of the northernmost Beja, the Ababda and the Besharin.

This was an area that required permits, but no one knew whether they were to be obtained from the Sudanese or the Egyptian side, as neither government had control of the region.

I decided to first try a formal approach. I enrolled at the University of Khartoum, thinking I would go into the area as part of an ethnological project. This was typical Western blindness on my part, because there was nothing formal there, no time for anything formal. Sudan itself was falling apart. At the time I arrived, strikes had paralyzed Khartoum and it was clear to most people that the government of Sadiq al Mahdi was about to fall. It was impossible even to obtain a permit to leave the city. Refugees were pouring in, from the war in the south, from the drought in Kordofan, and there were tent encampments all along the edges of town. The awlad a shams, "children of the sun," the thousands of homeless boys who had banded together, roamed the streets in search of money and food.

The University of Khartoum had always been an active political body. Only the finest young people in the country were admitted there. A Sudanese, wherever he or she may have come from, had to have absolute reading, writing, and speaking fluency in two completely different and very difficult languages, Arabic and English, beyond his or her own tribal language, even to be accepted as a student. Within the university there was a high level of discussion and commitment to solving the very real problems of life in Sudan. People risked imprisonment, even their lives, to act on their convictions, as did the professors who flew to Addis Ababa in 1989 to meet with Joseph Derang, the leader of the southern guerrilla movement, to discuss the possibility of peace between the north and south. They were arrested at the airport on their return to Khartoum.

Eventually I went back to chaotic yet predictable Egypt. I had been in and out of Egypt for a long time, knew the country well, and knew that in Egypt, which has, among other things, a very good road system, anything I wanted to do would be comparatively easy. I would simply aim for the necessary permissions, which I knew were nothing more than slips of paper stamped with an impressive circle of red ink that I could use to bluff my way through police checkpoints. If the permissions didn't come through I would go anyway.

The Crane grant had other complications. At precisely the time it came through (and I had waited through a year of interviews for the fellowship to materialize), for the first time in my adult life, I fell in love.

During my twenties, when I traveled a great deal, I became aware of a growing rift in myself. There was one side, and I identified it as male, that was eager to go into any kind of danger, and loved nothing more than to sit with a group of strangers, speaking a strange language, in the middle of nowhere. And there was another side, which I identified as female, that wanted only to stay home. I could hear their voices at different times, even at different times of day. The former I loved, for it longed for the rarest, most difficult stuff in life, the going beyond myself; but the latter, as the years went on, was becoming more emphatic, and more persuasive.

When I was awarded the grant, the conflict between these

two voices came to a head. Here was precisely what I wanted, a two-year license to wander around the eastern Sahara, and here was precisely what I wanted, the sweetness of intimacy, of having a real friend, of recovering feelings of trust and familiarity from childhood, and the possibility of finally having a home.

I went to Egypt as planned, and the man I had fallen in love with soon followed and asked me to marry him. Lanny and I were married in the fall of 1988, at Belhurst Castle on a cliff above Seneca Lake. Flo Whitwell had long since died, but my family still thought of Belhurst as part of the old eccentric Geneva to which she belonged. Flo had known the castle in her youth when it was a speakeasy (the speakeasy part of it hidden cleverly away somewhere up the grand oak staircase), and she had continued to think of the place as something rare and wonderful, the place for an occasion of any note, and so did we.

The marriage made me feel as though my family were mending again. My father, a judge, performed the ceremony. My little brother Charlie, soon to be married himself and now six feet tall, gave me away. I had called my mother from Cairo the month before and told her that I was getting married. This was something of a shock. She had long since given up on the idea that I ever would. We had tried to get married at the embassy, I explained, but it was too complicated and I thought the easiest thing to do would be to just go home. My mother rallied and within a few weeks pulled together a wedding that was a homespun, utterly lovely affair, helped along considerably by the generosity and enthusiasm of my old friends.

Jeanne Cameron filled the dark wood halls with red roses and lilies a peculiar shade of ivory she had searched the New York flower markets to find. Mike Hashim, who had moved to Geneva from Beirut when he was twelve (and gone on to become a jazz musician in New York with his own orchestra)played his alto sax, accompanied by an accordion (the only other instrument Belhurst would allow)—which was quite alarming to hear as I walked down the stairs to the altar on the arm of my brother. My bridesmaid, Georgina Owen, with our friend Tad Tharp, provided my wedding dress, an Edwardian tea gown they had gotten from an antique dealer in Washington. I wore it backward, showing off its icicle tiers of lace.

I arrived on the eve of the wedding, having come back from Cairo for five days for the ceremony. It was Halloween, the leaves turned but still holding, the water a hard grey-blue whipped white. For our brief honeymoon, Lanny and I stayed in the cottage on East Lake Road. It was closed up for the winter, and we had fires for warmth and candles for light. We read to each other aloud from *The Wind and the Willows* about the sweet familiar folly of Mr. Toad, and from Robert Lowell's translation of the *Oresteia*.

"An exile lives on empty hopes. I know. I was one."

Outside, Canada geese were settled on the icy water, the thousands of geese whose passage marks October in the Great Lakes. A glow of joy spread like the thick autumn light over all of this, over everything.

I WENT BACK TO CAIRO and set up a life for us there. This was toward the beginning of my time on the grant, and Lanny and I thought we might try to live in Cairo for a while. He was a writer, a journalist, and had written a great deal about the Middle East. He was not prepared for the Middle East that is Egypt, in which there is no news, in which there is only a sense of things slowly, dully, crumbling away.

An American friend of mine who grew up in Egypt said to me recently, somewhat wistfully, for she had been married and living in America for a long time: "You sit in Cairo and you see a bowl of fruit, slightly rotten and covered with flies, and yet the bowl, the fruit, the flies all seem to glow in their own soft light." It is not always easy to see this light. What you are more likely to see is the general atmosphere of squalor and decay.

I found a rooftop terraced apartment for us on Mohammed Mahzar Street in the embassy district in Zamalek, a wealthy neighborhood on an island in the middle of the city, where there was a thin veneer of Western life. We both got sick of it in a hurry. It was not Egypt, and it was not America either, and one felt there a tremendous sense of isolation from both.

The apartment did not last long. Early one morning I lay in bed beside the tall French doors that looked out over the Nile far below. The inescapable noise (nothing unusual for Cairo) of what seemed to be a secret, typically illegal construction project thudded away on the roof directly over my head. Then all at once the ceiling above me gave way and in its place I saw the grinning faces of the Saidi workmen who had been tossing bricks down onto my terrace for days.

I packed whatever I thought I would need into the back of my old Russian jeep, my Niva: my sleeping bag, what was left of the wedding presents we had brought—down pillows, monogrammed towels—cloth bags from the canvasmaker's full of books, a box of candles, a box of matches, a mag light.

I drove through the empty streets of the city before dawn, stiffly shifting the gears (which I had then barely learned how to use), and headed for the Eastern Desert, for the monasteries of St. Anthony and St. Paul.

Lanny had gone on to Jerusalem to write about the *intifada,* saying vaguely, sweetly, that he would be back as soon as he could. He had had enough of the life in Zamalek, of swimming at the Gezira Club, of sitting in the Zuhra Casino on the placid riverbank listening to the strained, furious anti-Western diatribes drone on from the mosques in Imbaba across the river. He thought it best if he went back to New York, where he had a job, a life, and waited out my time away; I could get out into the country and do what I had come for. I would regret it, he knew, if I simply gave up and went back to New York with him.

There began the endless pursuit of telephones, searching out the telegraph and police outposts that might have them, so that we could connect for three or six minutes, the minutes cut by sharp electronic bleats, the dear familiar voice pouring through the static like an elixir as I stood, sweating, caked with dirt, almost shouting to be understood, leaning at a counter in a crowd of soldiers, who would afterward, seeing my emotion, cluster closer and hand me cigarettes. (As one of them once explained to me, "When women are angry they cry, when men are angry they light a cigarette.")

Eventually we settled into something of a routine. Every few months we met in a hotel, usually a very good one, the American Colony in Jerusalem, the Pera Palace or the Aya Sophia Pensionlare in Istanbul. For a few days there would be the heady luxuries of lemons and ice and water and glycerine soap and the softness of the bed. My husband would leave in the dark for an early-morning flight back to New York. And I would wake up alone, almost ill with despair. Then I would straggle back to Cairo.

I come into Cairo disoriented, exhausted, and feeling like I should leave immediately. I stay. And gradually over the days I begin to remember the city. First I remember the words, finding them suddenly, involuntarily, forming in my mouth, on my tongue, like chocolate, delicious chocolate, "Shagal?" Does it work?

"Mahayashtagal." It will never work.

I walk under the Azhar flyway through the evening crowds and feel my way back, past the wet counters of juice stands thick with flies, the glass boxes filled with the greasy roasted insides of goats and lambs on rusted hooks, the pasteboard bookstalls between the jammed sidewalks and the old sinking mosques.

In Feeshaway's, dusty awnings cut the harsh light so that it filters down here and there, illuminating one thing, then another: heavy wood roses on mirrors with spiraling patterns of mold, brass acanthus leaves, smeared walls, fragments of long-dead animals, trash. Something in the shape of a star, a face, a hand. Frankincense and honey-soaked tobacco, stirring up memories, mingle with the smells of urine and sweat. A boy brings a glass of fresh mint, a clump of mint in a glass too hot to touch. He pours boiling water over it from a speckled tin pot on a brass tray where there is also a dish of sticky sugar, the grains slightly yellow as though dirty, flies sucking at the rim.

I walk back through the tourist bazaar, where odd things flash out from behind the glass: thick Sudanese snakeskins that look like the sleeves of old gowns, a thin malleable silver that glows with a soft yellow light, brass bowls filled with chips of lapis lazuli that can be gotten for a few dollars. Beyond the bazaar are the streets of craftsmen, divided by trade: tentmakers, perfume sellers, rug weavers, brassworkers, silver- and goldsmiths, men who make high stiff benches, which are also beds, out of the trunks of small trees. The mosques here are sunk far below street level. Water speckled bright green with algae fills a courtyard: the glass-green water of an open sewer, an oasis of life. The noise of the street intensifies with an undercurrent electric hum that is abruptly shattered by the evening cry from the Hussein Mosque. It echoes and expands through the air all around, a thousand times from a thousand mosques spiraling out through the city.

All the while I am walking toward the minaret of Ibn Tulun, the oldest mosque in Cairo. The minaret of Tulun has a staircase—the only such staircase in the city—that winds around the outside, suggesting, I think as I walk toward it, like its Arabic name halazawn, snail, the inner windings of a seashell, or the inner ear.

I walk up a mud street that winds through hills of garbage. On them children play amid dazed snarling dogs patched with mange. I enter the great door in the high crenellated wall that surrounds Tulun. Inside the wall is a moatlike space filled with loose dirt. The children surround me and ask for a pen, a piastre, and then kick up the dirt as I pass by them, head down, up the steep stairs to the great wood door, which is still open. The men are having their supper in the doorway of the old mosque. They offer me their food as I walk by. These are the men who watch the door: a clubfoot, a young walleyed peasant, a sharp-faced, neatly dressed old man. The mosque is closed now, but they let me in. They ask me where I have been. When will I come and pray with them, as I have said? Then they tell me the mosque is much too old. A holy place, but not a real mosque anymore. It is haunted. Am I not afraid to go back through the long dark corridors?

The cold stone is smooth beneath my bare feet. I walk away into the dark between the columns that stand three deep in the roofed portico. The huge empty courtyard contains shining light, a soft blue light. My whole being, at first, as I walk methodically around and around, feels like a limb that has gone to sleep. The chalk-white line of the Quran is carved in stucco around the upper walls, deep in shadow. A pair of *Athena noctua,* small white owls, fly overhead from column to column like angels in the dark.

The orange rim of the courtyard's rectangle of sky is marked on every side with star precincts: Sagittarius and Scorpio (Antares, its red heart of poison with its faint green twin; Shaula and Lesuth, the forked stinger in the tail raised to strike), the Dipper (Benetnash, the weeping girls in a funeral procession), and Cassiopeia (El Kaf, the open hand). The Citadel sits high on its distant cliff, a bright tin toy.

At the minbar the lettering, black and shiny, is tortoiseshell from the Red Sea. What is taken from the wilderness and set, gemlike, in the world. The quartz-flecked blue marble of the vine-crowned Coptic columns absorbed into the qiblah has been eaten by the air into waves.

A few hours later I have wandered back beneath the flyway to the edge of downtown and sit in the bar of the Windsor Hotel. The bar is a remnant from colonial days, with gazelle horns (shreds of skin still attached) mounted on its dark walls, dusty amber light, and red-and-white-striped satin cushions. Everything is old and worn. A television has been added since my last visit here three years ago, and *Falcon Crest* is just whinily beginning. Sweat forms on my forehead in the still heat. I am nursing a heavy gin and tonic, waiting for David to come and pick me up on his Jawa, a Czech motorcycle, small and cheap and fire-engine red, the vehicle of choice in this city.

I look forward to the ride. Perhaps he will take me along Salah Salem, the great swing of highway nearby, the road that runs beside the Mamluk tombs, where green bars of neon outline the pale mosques in the dark and the smoke of dung cookfires hangs like incense over the traffic. No. David is too practical. He will take me through some dismal part of town, because he wants to see it himself.

David is an old friend, one of the first people I call when I come into Cairo, although we may not have spoken for months or years. He is, by profession, an urban planner. Utterly American, as only an expatriate can be. He has lived in this part of the world for most of his life. He grew up in Beirut, and was sent away to school in America: Deerfield, Yale, Harvard. He came back to the Middle East as fast as he could. He was one of the last Americans to leave Beirut during the war.

David looks at the faltering sections of overwhelmed cities—Delhi, Kinshasa, Beirut, Kathmandu, Khartoum—where there is no longer any "infrastructure." The recommendations in his reports are purchased for vast sums, and he spends most of his time in Cairo, where he keeps an apartment for eighty dollars a month. Cairo is his favorite city. He claims he could not live anywhere else. To live here, he says, you must "get in the dirt mode," live close to the ground. Any attempt to shelter yourself leads to madness.

David comes at last. We make our way through the dense stalled traffic and into the back dirt alleys of the city. The ground is deeply rutted, a muck of dirt and leaked sewage. Peasants huddle around newspaper fires in the concrete shells of unfinished apartment buildings, their goats and sheep foraging through the garbage spread around them.

"There is no such thing as homelessness here," David says.

Women are dressed like princesses in long girdled satin gowns—pink, blue, green. Their headcloths glow white in the dark. They pick their way gracefully through the rank, broken streets, through the endless crush of cars and people, without looking down or stumbling. The crowds thicken around Saida Zeneb, Our Lady Zeneb, the great mosque. People are gathering for the Saturday-night hadra behind it at the shrine of Sidi Ali. Hadras are held in the tombs of dead saints every night. A hadra is a "presence," a dancing to bring down the presence of God.

An old man standing in front of a mechanic's shop in the dark garbage-strewn street leads us back to the famous shrine. We walk around the jewel-like, crumbling old tomb, a perfect angular Umayyad rosette set in its crown, and into the courtyard, where tiny bars of neon set off a purplish glow.

The wind plucks shriveled leaves from the dry sycamores.

Under them young men in black play screaming electric guitars. A blind ancient sheikh stands davening before them, chanting into a microphone. All around him people are dancing wildly or turning slowly or rocking to the music, eyes closed. Everyone dances alone, and at the same time with everyone else. The air vibrates. The dancers lift their hands as though to absorb the tangible atmosphere and spread it over themselves.

We are conspicuous foreigners in the crowd. We perch on the flimsy wood chairs under the courtyard trees. A tiny, fair-skinned man in an emerald turban brings us tea: the apotropaic gesture of courtesy. He wears a necklace of scallop shells. Children titter and run by, splashing us with perfume.

David whispers to me through the music, "Look at this and tell me about 'development.' "

In the early morning we are still up talking in his apartment, ashtrays overflowing on his stacks of paper on the floor. We are covering ground we have been over many times before. David talks, as always, with the tenderness of a brother, and he is unknowingly bringing me home.

"The A.I.D. agencies are a joke," he says. "Nobody gets it, what's going on in 'the Third World' these days. There are poets and novelists who convey the truth. But then, does one have to read fiction? Anthropologists only want to see remnants. Why are they always chasing after Bedouin? Why don't they look instead at . . . a slum in Bombay, say, or a block in Imbaba, to see how people survive? What is happening now is the real mystery, not the old stuff. What the West does not study, does not see, is that people here have made an enormous leap. Their resilience is a miracle."

He returns to the theme that has preoccupied him for years: what holds people together when everything around them is falling apart.

And he ends by laughing at me, how I tried to bring my husband into this mess, to set up an American life in Cairo. And how it caved in on me. How the roof fell down on my head. Well, what did I expect?

IL ADM BATUA GUA ISMU BISR"—The bone inside it is called a pit. "Now you go write it down."

Galal is sitting with Ahmed, his young nephew, and Um Naeema, the cook, eating dates, a clump of which he hands to me. I woke to their laughter and chatter and came down to find them all engaged in the weekly cooking. Ahmed is peeling potatoes. Galal is tasting things, harassing the others. Um Naeema is rolling grape leaves with her bare feet.

Her enormous black bulk bends the seat of her straw chair down to the metal plating of the deck. She is pale, Delta stock, with waist-long silver hair. Her shrill voice sings even when she talks. She choruses Galal, who is tiny beside her. There is an ongoing flirtation between them.

"What do you do out there in the desert?" she says to me.

Galal answers, although we have never talked about it: "She sees what the birds and animals are, who the Beni Adam are, that kind of thing."

"Oh," she says to him. "How's the weather in Aswan right now, hot?" ("Aswan" is the south, and to her, the desert, where he is from and she has never been.)

"Cold at night," Galal says. "In the daytime, from ten or so on, a furnace."

"Nice, you mean? Pleasant?"

"Oh, just delightful! Good for arthritis. People bury themselves in the sand up to their chins." Galal acts this out. "Then water pours off them. They put a little parasol over their heads. It's just wonderful. I hate to miss it."

Ahmed comes in the morning to work on the boat. Usually we meet each other in the kitchen over coffee. He has beautiful manners and a soft-featured, classically Nubian face. There is a nineteenth-century paterfamilias roundness to it. He has wavy black hair and a moustache. He removes his glasses and rolls up the sleeves of his frayed blue oxford shirt as he prepares to do whatever Galal erratically instructs—gardening on the bank, scrubbing the hopeless kitchen floor.

I make Turkish coffee. Ahmed drinks powdered Nescafé. He drinks it, he blinks to demonstrate, to wake himself up. He has been working all night as an electrician on city buses in a garage downtown. His father is dead and he has a mother and three sisters to support.

Galal catches a little orange-bellied fish and rushes it to the kitchen sink; it flops on the line as we applaud him. Afterward Um Naeema's daughter, in full Islamic dress, a starched white, conelike veil, comes out to fish. She is slight and girlish, a sight beside her mountainous mother, and has a gap between her two front teeth. She leans over the rail with her stick pole and line, a vision out of Maurice Sendak.

The bookbinder comes to take the books away. Galal carries them down in an open cardboard box and sets them on the side of the boat. We hold our breath. "I bet you thought I was going to dump them in the water, didn't you?" says Galal.

We decide what colors to have them bound in; the names pressed in in gold letters on each one. Om Sulaiman wraps the books up in the telephone wire that is strung up as clothesline on the lower deck, and carries them away.

He has left a box of newly bound books. I open *The Birds of Ancient Egypt,* and the smell of cloves comes out. I read:

> Flamingos on Gerzean redware, executed in silhouette only. It is as if the artists had intended to display these highly gregarious birds as they so often appear in nature, grouped together and forming what seems at a distance

to be a pinkish line of tall figures on the horizon. This decorative motif disappears at the beginning of the Dynastic era, and with this the flamingo all but vanishes from Egyptian art, only to survive as one of the standard hieroglyphic signs.

In a text from the New Kingdom, the ostrich is said to greet the dawn by "dancing" in the wadis in honor of the sun. Ostriches have actually been observed in the wild at sunrise, running around, spinning and flapping their wings, an activity which would indeed remind one of a kind of dance.

Evening falls and with it comes, as it sometimes does at this hour, a terrible loneliness. The voice of the Roda mosque shimmers over the river, streaked now with bleeding threads of green and gold. Low waves gently slap the metal below. A wedding on a distant boat plays a cheerful, metallic "Comet, it makes your mouth turn green, Comet, it tastes like Listerine . . ." There is the sharp whistle and crack of a cop shooting a stray dog in the street amid the pounding of car horns, then something closer, sharp heels on the deck outside. Imperious snaps at Galal, who comes back in abrupt, angry gutturals, "'Aiwa, Madame. Aiwa, Madame." Then, knocking, "Madame Shuzy! Um Adam." I open to find my friend Liz. Galal, muttering, immediately heads for the stairs.

Liz has just returned from a summer away and has a rumpled

paper bag in which there must be a bottle from the duty-free. Adam, her little blond son, stands behind her in the dark, staring away over the railing.

Liz, like David, has lived in Egypt for years, and it is hard for me to imagine being here without her. She has been married and divorced, founded a school, gotten a doctorate, made a movie, and learned the language very well. She is a small woman, but gives the impression of being quite tall. She marches undaunted into the offices of Egyptian bureaucrats, men's ceremonies in remote southern villages, embassies or the dull drunken parties of expatriates, wearing to all perhaps a Siwan wedding dress heavily sewn with coins, or a sunshine-yellow English skirt, deep delphinium-blue contacts, and faint lipstick on her pale, fine-boned Canadian face. She is from Kingston, Ontario, my mother's hometown.

We sit on deck with our legs propped up, sharing a plate of vine leaves and what little gossip there is. How is David? How is Niazi? Who have I seen? What am I going to do now?

I tell her I think I'm going to go home.

"You'll regret that," Liz says. "I tell you what. Why don't we go and see Radia?"

We will go see Radia together to find out what lies in store for us. Without really planning it, we make this pilgrimage to Radia, the Nubian prophetess, at the turn of every season.

Liz and I met in the Habu Hotel on the west bank at Luxor

in the winter of 1982. That year Liz was in the south collecting adid, women's laments. In the beginning this was a device, a way into the hidden, inner life of Egypt: the realm of women. Women, we knew, were the repository of the ancient traditions. The Habu temple, across from the hotel terrace, where we sat in the evenings, was marked with deep grooves where generations of village women had scraped the powdery pinkish stone away to mix into their medicinal potions for childbearing and disease.

After a death, men proceed with Islamic ritual, the Quranic recitation in the bright bare lights of patterned cotton tents. The women weep and daven together in dim back hidden rooms, sobbing out in choral response improvised poetic lines that go back five thousand years. The women do not sing of God and paradise, Liz told me, but utter despair. "We learned them in the school of weeping," they said.

In the laments death is a ravenous bird. Just like, Liz would say, in the Pyramid Texts, when Isis and her sister Nephthys mourn Osiris as kites circling above his dead body. The shrill whistling of kites is the first lament. Elegy in Greek, I knew, was a "whistling," ligu, the mournful whistling of carrion birds. Elegy is a metric form, a kind of love poetry that derives from something earlier, the funeral song, the love and longing that is close to grief.

We could pull the thread harder to unravel the thing, or at

least to get a sense of how eerily, how starkly, Egyptian religion derives from nature.

Osiris the god is a rotting corpse. Isis, the great mother, comes down on him as a vulture, a kite, the way a vulture descends on a corpse to eat it. The story is carved on an upstairs wall of the temple of Dendera: Isis fans the dead phallus up with her wings, and then descends on it, pulling life out of the corpse while devouring its flesh. She thus conceives the falcon, Horus, the living king.

The hieroglyph mut is the giant lappet-faced vulture, a creature seen only far out in the desert these days. Mut is the word for both mother and death, mother and mort.

In the beginning this was just another thread in Egypt, that when you pull it reveals the endless interweaving of old and new and spoken and seen. But then something happened that brought it all home. A few months before we met, Liz was driving on a road in the Western Desert with some friends. Beside her in the front seat of the car was an Egyptian writer, with whom she was deeply in love. They were chattering happily away on their way to a ceremony or a recitation in a village somewhere. As they rounded a sharp curve on the banked-up road, a tire popped from the heat of the asphalt and the Peugeot flew off into the sand.

Her lover had refused to put on a seat belt. He went through the windshield. He lay wrapped in blankets, unconscious, while they rushed through the nightmarish hours to get him back to Cairo, to a hospital. He bled to death in a traffic jam on the outskirts of the city.

When Liz and I first knew each other, our eyes would fill with tears as we talked about anything at all. We had many adventures in those days. There was the summer night I broke into her ground-floor apartment in Heliopolis, having just flown up from Dar es Salaam, and spread my sleeping bag on her front porch under a mango tree. I woke before dawn to find the poet Abnudi looking intently down at me. He had come, he said, to strangle Liz, but he would not, he said, if I would allow him to take me out to breakfast.

It was around this time that Liz met Radia, and later she brought me to see her.

Radia lives in Ain Shams, the "eye of the sun," where there is a large community of Nubians and Sudanese. The streets of Ain Shams are clumps of torn-up dirt. We have to pay the taxi driver extra to drive us the whole way. A row of open wood stalls bright with tangerines mark Radia's corner. Her building is the third up a narrow sloping side street. It is a modern concrete apartment building, five stories high, fronted with stucco. The bottom floors are unfinished, and there is no light as we walk in, up the uneven, worn stairs, seeing dimly back through the naked beams into a mess of bottles, melon rinds, newspapers, the fur crust of a dead cat.

As we arrive at Radia's landing on the third floor, the light snaps on and the door opens. She is not at home. Her son, a jazz musician in his late twenties, lets us in. Radia is a matriarch, and three of her sons live here with their wives and children, as well as her unmarried daughters.

The apartment is heavily decorated with bright synthetic Indian rugs. The furniture is stiff imitation Louis Quatorze, painted gold. An enlarged photograph of Radia's dead husband posing with a sheikh, bordered in pink, hangs in a gilt frame on the wall, beside Hallmark photographs of dogs and cats. *Batman* is playing on the VCR with Arabic subtitles. Radia's newest grandchild, a girl of three, sits burbling on the floor before the set, where Adam joins her. We hear the chanting from a nearby mosque.

Radia comes at last. The ceremony is always the same. Her daughter brings us coffee in tiny white cups rimmed with gold. We drink slowly and talk lightly about this and that, people we know, Adam's school. When we are finished, Radia takes the cups from us and with a firm flick of the wrist flips them over into their saucers. So they sit for ten minutes, until the webbing of the sludge dries on the white porcelain sides of the cups. Radia then picks them up, and out of the webbing reads to us about our lives.

Our friend Gamal laughs at Liz's faith in Radia, and in others like her. But I shuddered the first time she looked up at me in alarm with some precise detail from my past that even Liz did not know.

Today she looks at me intently, her yellow headscarf slipping back, her bright, playful face suddenly hardening, looking very old. "You must be careful not to be torn apart," she says. "You must be careful to limit yourself, or conflicting things within you will tear you apart. This is what leads to mental illness. Your qarin"—here Liz looks at me; we have talked about this, but it is sihr, secret, and I have never heard it spoken of as a real thing—"your qarin is angry, choking you from within because you will not do what he wants. This is the tightness you feel in your chest."

My qarin is my male twin, my invisible brother, my shadow soul.

"Leave me an article of clothing," Radia continues, "just a sock will do, and I will bring it to a certain Sudanese I know, a magician. We will use it to call up your qarin. We will see what he wants, how to pacify him. Come back to me the day after tomorrow, and we will proceed from there."

—

In the Cairo Museum, in the upstairs room where desert flints are displayed: impressive, huge, prehistoric axes and knives knicked evenly all over their hard glassy surfaces like wind running across the surface of a wave. Flint is natural glass—melted, metamorphosed sand. Below them are file cards on which is typed: "From the Great Sand Sea."

A young Egyptian couple walk beside me. The man says, "Could you tell us, please, what is the Great Sand Sea [the Bahr el Ramla]? We have heard of the Red Sea, we have heard of the White Sea [the Mediterranean], but the Sand Sea? What is it? Where is it?"

Downstairs I stand awhile before the nature gods of Sahure. They are standing sideways in a line as though taking a casual walk together. The relief is carved with the simplicity and soft precise lines of the Old Kingdom. There is a god whose body is grain, his skin incised with oblong flecks of yellow. Another's body is made entirely of waves, line after line of jagged w's, three of which together make up the hieroglyph for water. There is a fat yellow goddess, Peace, and a goddess whose slender young body is green from head to foot. This is ib aw, joy, "the opening of the heart."

On my left index finger Radia places a silver ring marked with signs. Some of them are Coptic letters, others I do not recognize.

They look like Phoenician characters, jagged lines that could be a human figure, crowned with horns.

Over the last few days Radia has given me packets of incense wrapped in lined paper tied up with string. Drawn on the paper are similar signs. My instructions in this, the exorcism of my qarin, have been to bathe myself in the smoke of the incense in the morning and evening of each day. I place each packet from the progression in a bowl of smoldering charcoal to release the smoke. Charcoal is the flesh of a tree, Radia explained to me, and so of the earth as I am, kin, and necessary for help in the ritual. There can be no direct use of fire. I cannot let water touch my skin during these days.

"Now you are free," she said to me today. "We have pacified your qarin. The way is open before you, and you can go."

I took a taxi back to the boat. The driver was a sharp-faced young man. An Egyptian girl sat in the front seat beside him, heavily made up with blue mascara and eye shadow with gold sparkles, her short hair dyed patchily blond. They shared a pack of cigarettes during the long ride and were so friendly I could not tell whether they were in love, engaged, or whether she had just gotten in as I had.

The driver was dreamy and talkative. He recited poetry. He sang. He said I was so nice I was a sherbet. "Inti sharbut."

"What?"

"You know, like out of strawberries."

The girl put her window down, and he said her hair would be "muhafhaf," disarranged by the wind. "We should all be in helicopters," he said. "All people want these days are cars, apartments, television sets. They are willing to work so hard, and they finally get what they want. Why, look at this delightful collection around us. What extravagance! What expense! What a mess."

We were stuck for a while behind a car at a corner on the Giza corniche. The car had a red crescent on the back; its driver was a doctor. The doctor was a jaban, a coward, our driver said. He was afraid to turn into the oncoming traffic. Our driver abruptly got out and walked into the street along the corniche. He began directing traffic, stopping the endless stream of cars with his hands so that the doctor could make his turn.

When I got out at last, he said, "You have honored us," and refused to accept the fare until I folded it and placed it on the seat beside him as a gift.

THE FALL MIGRATION:

Green bee-eaters are tossing above the boat, barking like dogs, barking for joy. Their sheeny electric-green bodies fall rapidly toward the water. Just before they slam into it, though, the little birds pull themselves up again, their long black tails fluttering out behind. By the sound there may be twelve birds together. I rush to the door to watch.

As I open it, something falls wriggling and twisting to my feet, curling in rapidly, painfully, upon itself. It continues to move in this way for a long time. Too long. At first I think it is a large flesh-colored worm. It has faint black stripes and is almost translucent. It must have been resting between the door and the frame above, and I must have crushed it, killed it, with my sudden movement. At last it stops moving, must be dead. I crouch down to see what it is: a gecko's tail. In an hour the ants have carried it away.

The wind covers the boat with dust, slamming the glass windows open and shut. The vases are filled with dead roses and mosquitoes. There is a thick coating of dust on everything, so I leave footprints wherever I go.

In the night I wake to the wind scraping against the door, gasping through the shutter slats, becoming shrill, falling, then rising again. I imagine the demons that Radia has cast out of me are scratching on the walls outside, trying to get in. But they cannot. I have my ring.

In Niels Bohr's model of the atom, energy is emitted as an atom passes from one state to another. This energy is color. I could think of color in terms of words, in which color equals cover, for they stem from the same root and contain the same essential idea: the veil. As Saad used to say, "Ya Sater, ya Rabb"—O Lord, O Great Veil. And I could think of color in terms of what happened to me when I went to Egypt. For me going to Egypt was a passage from one state to another. It involved three polarities, three passages between extremes.

The first was my transition from feminine to masculine. In Egypt, I became a man. A category already existed for this condition: foreign woman as honorary man in Muslim society. Many women had taken refuge in it before me (and, I am sure, were drawn to the Middle East precisely for this reason). A solitary Western woman does not fit into the realm of women, which is the family. Indeed she is often prevented from entering it altogether, in physical terms, from entering the inner quarters of a house. She is received in the forecourt by the men. At the same time, a Western woman does not have the threatening presence that a man does. She has the best of both worlds. She is at once a stranger, and so treated with the respect due a stranger, and as a woman alone is a person in need of protection who inevitably becomes a familiar.

I had learned how to enter this state of honorary man quickly, with a stream of guttural Arabic, the last thing anyone expected to hear from a young American woman. And I entered it joyfully, for here was another self.

I once saw how much language was a part of this transition. I was in the qawa in the Wadi Baramia with Saad and Jesus the tea man, who was making us tea. The qawa was a welcome stop in the shade at midday on the long desert road, a dusty shack of reeds and concrete, and as usual there, the three of us bantered cheerfully away. Saad told Jesus that I was a spy, and that he was showing me all the places in the desert where there was hidden gold. "It's true," he said. "She's an American, you know. That's what they're after, gold. Show him you're an American.

Say something in your own language." And I began to speak in English for the first time in weeks. I was struck not only by the look of surprise on their faces (for they had never seen me in this mode), but by the sound of my own voice, how high and soft it was. With the voice I took on my old identity. I was a woman again.

The second polarity that drew me to Egypt was the passing of nature into language. I had been keenly aware of it all my life.

The translation of nature had to do with the eye, I thought. A mystery surrounded the question of why the hieroglyphic sign the eye was the verb to make, rather than simply to see. It reminded me of something that Emerson wrote:

> The poet is the Namer or Language-maker, naming things sometimes after their appearance, sometimes after their essence, and giving to every one its own name and not another's. The poets made all the words, and therefore language is the archives of history, and, if we must say it, a sort of tomb of the Muses. For though the origin of most of our words is forgotten, each word was at first a stroke of genius, and obtained currency because for the moment it symbolized the world to the first speaker and to the hearer. The etymologist finds the deadest word to have been once a brilliant picture. . . . Language is fossil poetry. As the limestone of the continent consists of infinite masses of shells

> of animalcules, so language is made up of images or tropes, which now, in their secondary use, have long ceased to remind us of their poetic origin. But the poet names the thing because he sees it. . . .

The third polarity I understood in my experience in Egypt was between the city and the desert. Cairo was Um a Dunya, Mother of the World, teeming with life made possible simply by the presence of water. In Western terms it was a city that did not work. It was overwhelmed with people who lived under conditions of privation and crowding greater than any before known in urban life. As my friend David used to say, these Third World cities where millions and millions of people are crushed in together are a completely new phenomenon on the earth. There seems to be no structure in them. Yet this was, as David said, precisely where the mystery lay. The city did work. There was a structure, but it was hidden, internal, because the people in such cities were adaptable, resourceful, compassionate—in short, alive. In Cairo they got by with remarkable cheerfulness. Cairo was a nonviolent city, with comparatively little crime.

If the city was Um a Dunya, what was the desert? The city's negative: a blank page on which things magically appeared. Again it had to do with color and the eye.

The desert's pure air was a prism separating the harsh light into bands of intensely articulated color. Vision there became a joy. The eye in the desert could see with such clarity that something as subtle as an animal track, which in the city would be lost, stood out distinctly.

I have sometimes thought that this quality of light in the Egyptian desert is what inspired the monasticism of the early Christians and the Sufis, who spoke of an intoxication of the spirit there. To see it is enough, is like falling in love.

"Traditions differ respecting the fabric of the seven heavens," begins a note in Edward Lane. "In the most credible account, according to a celebrated historian, the first is described as formed of emerald; the second, of white silver, the third, of large white pearls; the fourth, of ruby; the fifth, of red gold; the sixth, of yellow jacinth; and the seventh of shining light." That progression of color is a fair description of dawn in the Egyptian desert. The light itself is precious beyond all wealth.

WHERE THE CRANES LANDED

IN RAMSES STATION THE AIR IS DENSE with noise, voices, movement, the heavy undercurrent throb of the engines of standing trains.

Shafts of evening light thick with dust slant down from the high windows into the dirt-encrusted halls. Lean dark peasants in pastel gowns carry urban plunder on their heads: ghetto blasters, sewing machines. Soldiers are everywhere, either posted with Kalashnikovs around the station or off duty and heading for their distant village homes in third-class night cars. Gnarled old porters in blue slip silently through the crowd like fish in dark confusing waters. A group of pink tourists stands frozen as though shrinking together against the stream.

I walk out onto the platform and, for a moment, mistake branches waving in the distance for heads bobbing over a bridge.

In central Cairo, the eye expects to see people. But now palm trees move behind the rusted-out shells of trains. A brownish-gold haze rises over the city upward into a lavender sky. By the tracks an infant girl, smeared with dirt, lies screaming on the concrete beside her father, who has made a couch for himself out of twine-bound boxes and rugs, and leans back, contentedly smoking and drinking tea.

I sit on my duffel bag of blue-and-white-striped cotton against the yellowed marble of the wall, sweating, sore feet in boots, a small bottle of water beside me. I am waiting to board the *qatter al nom,* the overnight train to the south. I am on my way to see my old friend Gamal Sharif to ask him if he will accompany me out to the Red Sea. I need to find a host among the nomads there, so that I can travel and stay awhile in their territory: the desert and seacoast south of Quseir. Even if I do not find such a friend among the Ababda on this trip, I know it will be good to see Gamal. We have made many excursions together over the years.

As the train leaves the station, men in rags scramble up to ride between the cars. When we reach the outskirts of Shubra, all of a sudden, Cairo looks eerily like New York, in the Harlem–Yonkers stretch of track. But here the buildings are not abandoned but unfinished, and at the same time moving toward collapse, as if they had passed from a building's equivalent of childhood to old age, without the intervening stage. Banana trees and date palms grow in the trashed-out lots between them. Pigeons swirl overhead: every one of them is owned by someone and may someday be the owner's supper.

The train rocks violently on the tracks. An old Sudanese woman sits on the berth below me. She is small, very dry, with cheeks like withered fruit, and with three blue dots on her chin. Her lips are also blue. Her grandchildren are screaming. When I came into the compartment, she (after the initial shock of seeing me there) pulled down the upper berth, a slab of wood and foam encased in red plastic, and motioned me to get up on it. I did so, offering fervent greetings and protestations. Some of her children, who live in Cairo, had come to see her off and consoled her for having a kafira in the cabin.

When I am a safe distance away, cross-legged on the berth with my neck curled down to duck the ceiling and my back rocking rhythmically against the side of the car, she looks up from where she sits on the floor arranging a dinner of french fries and pink pickled turnips for her grandchildren, silent now at the sight of food. She motions me down to join them. I will never starve among Muslims. French fries are my favorite food in this country. But I decline.

The conductor enters. He yells at the woman for sitting on

the floor. She pulls her black sheet snugly around her chin and says nothing. He apologizes to me in English and says he will find me another compartment.

When he returns, the grandchildren are asleep at the foot of the lower berth and the woman, still on the floor, is braiding up her long silver hair. I tell him I am content to stay where I am, and he leaves. The old woman smells powerfully of tobacco and wood smoke. I drift off to sleep.

I wake in Upper Egypt to a silver-blue sky and see a wild turbulence of plants: palm trees, corn fields, high tufted reeds, all against pale distant cliffs. The corn-husk hovels of farmers stand amid vivid clumps of sugar cane with raw white stems. Even in the train I can smell the air of Upper Egypt, the vanilla smell of the bean fields, the pervasive, dry sweetness of the acacia trees, a smell as distinctive and sweet as that of locust trees or milkweed.

I have taken this train a number of times over the last nine years. I used to think of it as the true journey into Egypt, the Egypt of the Said, the high country. It felt like going back in time. I traveled third-class. The farmers would hand wands of the sugar cane they were cutting through the open windows. They were stacking the cane into carts behind their water buffaloes—exhausting work—and yet they would stop and cheerfully give some of their harvest away.

One of my first impressions of this side of Egypt, far from Cairo, was its slowness. In the oases, in the far-flung villages of the Nile Valley, many people had never had the experience of speed, had never been in a car or on a train. They thought of time in terms of distance walked or the actual, uneven growth of a plant or a child.

We arrive in Luxor at last. Luxor is one of the seedy towns in the world where Eastern culture meets Western in the worst possible way, as something to be bought. The transaction breeds an atmosphere of prostitution and contempt. I head straight for the ferry and the west side of the river. I will have to see if the Niva, my Russian jeep, will start after having been parked in the desert for three months.

I used to be afraid of cars. I never really learned how to drive before I bought the Niva. I bought it in Cairo from a French engineer who was on a two-year contract to build the Cairo metro. He hadn't used it very much, and had never taken it out of town. I didn't realize when I bought the car that it would be against the law to sell it again. "Just push it in the river when you leave," Liz said. "Easiest way to get rid of it." She had done this once with one of hers.

In the beginning I was uneasy about the responsibility of owning something as solid and demanding as a car, and, more than this, something that involved serious paperwork.

I understood that if I hit someone—and how could I miss in the streets of Cairo?—the fault would be entirely mine and I could end up in an Egyptian jail. The Niva sat in the garage of the building on Mohammed Mahzar Street in Zamalek. Lanny and I went out to Saqqara one day, and he taught me how to shift the gears as we rolled slowly around the collapsed pyramids in the sand.

When the roof of the apartment caved in and I found myself without a place to live, the Niva was waiting downstairs. Just before dawn I drove slowly through the empty streets of Cairo, knowing I would only attempt such a thing at this hour. I drove toward the desert and the Suez road, then on to the Red Sea coast and the Zone of Dangerous Curves.

The Red Sea road was a fresh, soft run of new asphalt. There was never anyone on it, only the wind to help you along if you were going south, and slow you down, and buffet you gently from side to side as you drove north.

There were six hundred kilometers or more of open road. I learned how to shift evenly on the long, empty stretches, pulling off the road every few hours to raise the hood and let the engine cool down. I drove as far as I could, to the village of Abu Simbel near the Sudan border. I drove for three days, stopping only at night when there is danger because of the truckers who drove without their lights on.

The Niva broke down in Safaga, the port where American grain shipments come in. I sat for a day with the mechanic's wife and sisters on the floor of their concrete house watching a Fred Astaire movie on TV and eating cheap hard candy as children and flies crawled all over us.

Since that time the ministrations of mechanics have been a regular part of my life in Egypt. The engine, ever filling with dust and sand, needs cleaning, needs parts replaced—a difficult prospect after Egypt stopped importing these cars, when their relationship with Russia soured. I know the dealers of used-car parts on the back streets of Cairo and Luxor and Kom Ombo, and bargain with them for objects that look as if they have lain neglected in someone's basement for fifty years.

The Niva is good in the desert. It is light, unlike, say, a Land-Rover or a Mercedes jeep; you can push it when it gets stuck in the sand. When it overheats (as it frequently does) you simply turn it into the wind. I finally got to the point where I could race it over dunes—an odd, slipping sensation, something like ice-skating. I began to love the car.

It is a joy to have the Niva back, with the dusty, burnt smell of its black sun-scorched seat and its gear box stiff from disuse. I drive south to test it out along the narrow west-bank road. Just below Armant, I round a bend on the canal and see above the broad green flood plain of the Nile a hundred or more great white storks circling into a vortex that reaches high up into the sky. The huge birds fly 'round and 'round one another, passing one another silently as they rise.

I ARRIVE AT GAMAL'S TOBACCO STAND by the Luxor train station in the evening. He is sitting with a young Swiss woman from the Oriental Institute in Paris. They are translating Sufi poetry on sheets of paper spread out on upturned cardboard boxes on the street beside his doorway. Gamal spent hours the night before with her Walkman on his ears, transliterating the tapes she collected in Qurna this week.

The woman is a jewel in the filthy street, and people stare at her. Her lips—without lipstick—are almost scarlet. White-blond curls stick out from beneath turban cloth wrapped bandana-like around her head. She has a clear, strikingly lovely face: slightly waiflike, but with the distinctive sharpness of a woman who is used to getting around alone.

Her name is Rosline. She tells me she began to study Arabic

when she met a man while backpacking in Mexico who told her, "Learn Arabic. You will really enjoy it."

She leaves, and Gamal and I are alone. I suggest as casually as possible that he and I take another trip over to the Red Sea. He listens to the idea in silence, his back turned as he assiduously empties newly arrived crates of Pepsi and Fanta into the freezer.

Gamal hates the desert. He even fears it, as a city person does. The desert is as wild, unpredictable, and dangerous as the open sea.

But, after all, we are old friends, and he has been tolerating this desert interest of mine, as he tolerates the projects of others who come to him for help.

What Gamal really knows about are words, poetry, the oral tradition in the south. There is a local joke that he has several doctorates, none of them his own. He walks foreign researchers through their fieldwork, introducing them to relevant sources, helping them collect and then interpret material.

Gamal does not advertise this service, which benefits him not at all. His name is passed along from one researcher to another. He helps them through an obscure sense of honor, simply because he is asked. Once in a while he will get a little money at the end, or be mentioned in a footnote somewhere.

These excursions, he will say, help him with his own project. For years he has been working on a dictionary of the Saidi dialect, which has a large infusion of obscure words, words that have never been written down.

Southern Egypt is heavily Christian. The Copts fled to the south under Roman persecution and were isolated here, living in temples and desert caves. Their language, the link to Ancient Egyptian with Greek influence, is still spoken in places, and Saidi, which the peasants speak, is full of Coptic words, fragments of the earliest language, often simply double sounds.

"You collect words like butterflies," I say to him.

"Yes," he says. "I go from flower to flower, taking a little from every place."

"No, I mean the words themselves are the butterflies—and you collect them."

"I see," he says judiciously. "Well, both meanings will do."

Today I am only asking for his company.

I go to Gamal when I need a friend. I can sit with him for hours and say nothing, in a physical empathy in discomfort, sharing an edge of shade in the intense heat, smoking cigarettes, drinking hot tea, and leave with the feeling that I have confided in him all the secrets and difficulties of my life and received a deep, nourishing sympathy. There is a kind of sensuality that is distinctly Egyptian, a surrender that carries one through the harshness of life to the point of exhaustion and then beyond it, to something like relief.

Gamal has a lean cowboy look, hollow cheeks, a shock of silver-white hair. He is a distinguished figure in Luxor, a muhandes—a college-educated man and, more important, a Hagagi, a descendant of the city's patron saint, Abul Hagag. Gamal worked as a water engineer on the Aswan Dam in the sixties and still has this title, spending his days over papers in an office in town. But he cannot feed his four children on the government salary of forty dollars a month. The real money comes from the "duchan," as he calls it, accent harsh on the last syllable, like a too familiar, unpleasant thing that must be named: the tobacco stand, which he keeps open all night.

The duchan is between the Nubian Café and the Bazaar Sudan. The owner of the café is throwing buckets of water on the dusty sidewalk so he can set out more tables, burning frankincense on charcoal to cleanse the place of spirits.

Madmen collect on this corner in front of the railway station. Tonight a small fine-boned man with thick dark hair and white stubble on his chin walks back and forth in the gutter, mumbling to himself. He wears a sky-blue gallibyah, torn up the side and blackened with soot. A second gallibyah is thrown around his neck as a cloak.

The Nubian sweeps dust over a boy's stack of evening papers as he scrambles to pack them up and leave, awkwardly. His left hand is paralyzed. The madman cries, distraught, "How can you do this? Are we not all human beings?" The Nubian throws a bucket of water at him, drenching his clothes. Enraged, the man leaps on the sidewalk brandishing a palm stick. The Nubian tears it out of his hand and beats him with it, and douses him again. The men sitting at the café tables laugh.

Gamal drapes his bony arms about the man and leads him away, gently feeding him a cigarette. Gamal is disconcerted. He has just been telling me the proverb that a Nubian, though his face is black, has a white heart. (Gamal himself has very dark skin, Nubian blood. His brother's name is Nuby, "black.")

As we split a ritual Coke, two German backpackers walk up and point to a pack of Cleopatra cigarettes.

"How much?" one of them asks.

"Eighty-five piastres, thirty cents," Gamal says, in his slow, careful English.

"You lie." They stomp off.

"But it's a government-regulated price." Gamal says as much to the air as to me, spreading his hands in irritation and wonder. "The police would close me down if I didn't post the government price." But he is used to this, and ends with his bittersweet laugh as he comes up with a new proverb: "Once we had wars, now we have tourism."

Ragged Egyptians appear and buy their cigarettes one by one. Children, squeaky clean, with ribbons and shined shoes run up to buy penny candy, or squares of honey-soaked tobacco for their fathers, who follow them up with, "Well, Sharif, what's the news?" Gamal has a little red TV set back on his freezer, a gift from an American friend, "Suzanne tanya," the other Susan, he sighs, who has long since become a professor somewhere and disappeared.

"Do you know how hot it is in the desert at this time of year?" he says at last. "You realize we could die out there? We will have to carry all our own water and gasoline."

I understand now that Gamal, in a sidelong and reluctant way, has agreed to come.

There is a good reason for him to make the trip. If we go south we can make the pilgrimage to Hassan a Shazli, the tomb of the Sufi saint who died on the old Mecca route through the Red Sea Hills in the twelfth century. In southern Egypt people believe that making the pilgrimage to the tomb of Shazli three times in a lifetime is the equivalent of going to Mecca. Gamal does not actually believe this, but he is a pious man. He knows he may never have enough money to go to Mecca, and he has been wanting to make the pilgrimage to Shazli for a long time. It is the best he can do.

We decide that this will be our ultimate goal. Shazli is south and inland from Marsa Alam. We agree to set off for the geological survey station at Marsa Alam in a couple of days. Then we will head south from there. We begin to organize our provisions with things from the duchan, matches, tuna fish, and bully beef.

As we look over the dusty shelves, a small, pale man appears outside, so quietly and gently that he seems almost like a ghost. It is Rizq the weaver. Rizq has the round, deep-sunken eyes of an infant; his thin, delicate skull is visible through the papery skin of his face.

Gamal brought me once to hear Rizq sing the ballad of Mari Ghergis, St. George. St. George had always been close to my heart. I used to carry a postcard icon of the saint. He was, Gamal had long since explained to me, the Christian version of El Khider, the Green Man, who appears to travelers in distress. Gamal had taught me the passage from the Quran which begins

"Once Moses was walking along with his servant boy, and he said, 'I won't stop walking until I find the majmual bahrein, the juncture of the two seas, even if it takes me the rest of my life.' "

The boy was carrying a fish. When they walked past the juncture of the two seas, Moses did not notice the place, but the fish slipped out of the boy's hands and swam away. This is when El Khider appears to Moses.

"What about this juncture of the two seas?" I would say to Gamal. "Are we talking about the tip of Sinai?"

"There are two seas in the world," Gamal would say, "freshwater and salt. Two completely different things, one gives life, the other takes it away. And the majmual bahrein is where they meet."

St. George is the most popular saint in Egypt. Copts and Muslims alike come on a yearly pilgrimage to his tomb at Armant for his mulid, his birthday festival. There is a thought that St. George is distinctly Egyptian, is Isis spearing her evil brother Seth, who has turned himself into a hippopotamus, and that the monster under the water is the rising of the water itself, the seasonal flood of the Nile.

Rizq says that once when he was very ill he prayed to God to see St. George. He prayed and prayed. Late at night the saint came to him, removed the blockage from his side, and threw it into the river. The saint then washed his blood away with Nile water, and in the morning Rizq was cured, without money or a doctor.

Rizq is a Copt from Negada, a village not far from Luxor. Negada is an ancient weaving center. In his family house, a mudbrick house from the eleventh century, is a wooden loom pressed into the mud floor. His aged sisters weave threads on the loom into a deep pink silk striped with gold, with black pyramidal designs along its edges. Sheets of this cloth become wedding gowns for Sudanese Muslims. The monks in the Coptic monasteries near Negada long ago took the pattern from the illustrations in pharaonic tombs, Rizq says. Negada is a predominantly Christian town. Although Gamal is descended from a Muslim saint and Rizq is from an old Coptic family, they have always been friends.

Tonight Gamal says to Rizq, "Do you have your book with you? We are going on a journey."

Rizq reaches into the pocket of his threadbare gown and pulls out a hand-sized book. It is worn and bent and has a thick paper cover marked with signs in red ink. Gamal handles it with great delicacy. "This book," he says to me, "is a very ancient book of spells from the monks in Negada. Perhaps Rizq will write us out a spell for our protection."

"Do you have a spell for protection against scorpions?" I ask him.

"There are sixty-six spells against scorpions in this book." Rizq leafs through it, finding a short paragraph that begins with a line of words written in red, as a paragraph in hieratic on a papyrus will begin.

"And for money, for success in business?" Gamal asks.

"I have been thinking of precisely that. I have been meaning to copy one out for your shop."

Rizq comes into the shadow of the doorway and, opening the book, scrawls out two lines on a torn scrap of paper with Gamal's Bic pen. Gamal examines it for a moment, then tucks it beneath the cardboard that covers the waist-high cabinet in his shop where he keeps Lux soap, candy bars, and cans of tuna fish. Rizq continues to write, then hands me another scrap.

"This is for protection on your journey," Rizq says. "Matka-feesh." You need not be afraid.

As he closes the book to put it back in his pocket, a group of boys walk by, a little too close to the shop. They mutter insults and quickly shove into Rizq from behind. Gamal shouts at them, "Aeeb!"—For shame—and they scatter. "I don't know what's happening here," he says after Rizq has left. "Egypt has always been a tolerant place."

I come back to the mudbrick house on the west side of the river, where I sleep out on the open roof. I find its owner, an old Saidi farmer named Om Salim, standing in the dark with an Enfield rifle so old that it looks like a toy. He says he is guarding his sheep from the wolves that come down from the cliffs at night, and then there is the itf, the swamp lynx, rustling in the rows of sugar cane beyond the house. Can I not hear it?

On the roof above I lie out on the cool nylon cover of my sleeping bag as the moon climbs through the palm trees in a haze of greenish light. As it emerges, a violent wind comes up. For a long time I huddle down in the wind, feeling it parching my face. It comes in gusts, tearing at my hair and at the sleeping bag beneath me.

I go below and take shelter in the kitchen. Moonlight streaks in through a high window. A pale, mottled blue scorpion scuttles into its buttermilk path on the floor. I crush it with my foot. The red ants swarm out through the cracks in the tiles and eat its blood.

—

"You must bless everything that happens," Gamal says, "so that more of the right thing will happen."

He arrives before dawn at the ferry landing on the west bank. He carries a briefcase, and is dressed for work in a frayed blue buttondown oxford shirt and brown trousers.

We provisioned ourselves the night before with jerry cans full of gasoline and water, and with canned food from the duchan, tuna fish and bully beef and a jar of Fayyum honey tasting of oranges.

So we set off. We drive down the west side of the river, crossing to the east at Qesna. We both feel the excitement, and fear, of not knowing where we are going, where we will sleep that night, what driving out into the desert will be like. We don't talk about it. We head for the Edfu–Marsa Alam road.

There are three paved roads between the Nile Valley and the Red Sea. They follow the routes through the dominant wadi beds, the dry watercourses that once were a network of rivers feeding the Nile. The southernmost runs between Edfu and Marsa Alam. There is a break in the limestone cliffs along the east side of the Nile at Edfu where the wind has eaten the limestone into ribbed islands pocked with caves. The break in the rock makes a door into the Eastern Desert which it otherwise holds back, as though behind a wall.

The temple of El Kab is here, the precinct of Mut, the great mother as the lappet-faced vulture. The temple has decayed and been rebuilt many times in this place since prehistory. Five miles south along the river is the temple to Horus, the falcon, her child.

Just north of El Kab the valley narrows and is lined with high ridged cliffs. Mudbrick houses are built up against them, tall and narrow like the castle towers of the Hijaz. We stop for a breakfast of beans cooked in water buffalo butter with handfuls of fresh arugula and hard, salty cheese at a little roadside stand here, a building made of mud with latticed sides. A few truckers lie on long benches in the palm-thatch shade, resting up for the harrowing ride on the narrow valley road. Between here and Aswan, a hundred kilometers to the south, the road is full of curves and blind, and many people die on it.

"You must bless everything that happens so that more of the right thing will happen," Gamal says again, hollow-eyed. He has been telling me about the stress of living in Egypt these days. What will become of his children? "Egypt is a boiling pot. Count on it, it is about to boil over."

We drive away, east, directly into the harsh morning sun.

THEY ARE REPAVING THE DISINTEGRATING ROAD between Edfu and Marsa Alam. Tarmac is just another rock in the desert. A road will last about five years before breaking up into rough black clumps. We hit them with a thud. A dumptruck full of sand almost backs into us. Gamal shouts: "Calax! Calax!" Honk! Honk!

I do so. The driver stops and waves, smiling benignly.

I slalom the Niva around piles of rocks littering the road. Sand swirls up in tall funnels in the wind. We drive straight through them. The sun makes everything scalding to the touch—the car metal, the black wheel in my hands, the black plastic seats.

Out of the Nile Valley the first sight: a hundred camels pressing up against thin brown acacias. A surveyor, plump, urban, face wrapped completely in a white scarf save for his eyes behind thick sunglasses, sits under a flowered pink umbrella. Another struggles with a theodolite up the road.

I rush through the police post, as I usually do, with a foreigner's blithe sense of immunity. But this time the soldiers have set up an oil barrel in the middle of the road. I do not quite miss it as I intend. The barrel spins over on its side. Gamal leaps out of the car and diffidently picks it up, and brushes it off as though it were a valuable piece of machinery.

"She didn't know she was supposed to stop," he explains to the three young soldiers who have come out of their concrete bunker, drowsily, with guns. Gamal puts cigarettes in their mouths and lights them as the soldiers look him up and down and I glare at them. "They don't put barrels in the road in her country."

We leave the checkpoint behind, and I see that Gamal is shaken. "Look," he tells me when we are a good distance away, "this is the border. You know what that means."

"Drug smuggling. Arms smuggling."

"Cops. Everywhere."

We distract ourselves, as we often do, with language. We talk about the word dragoman—a role Gamal often plays. He tells me the word comes from the pre-Arabic bergil, to draw a square, which also means to evade a question, or to be afraid. No, he says, it comes from bergim, the language of doves, because a foreign language sounds like the language of birds. I tell him that in English: dragoman has a medieval quality and sounds like "dragon man," snake, a clever person, but with hidden ways. We decide in the end that it comes from a word he and I often say to each other, tergim: translate.

The word has a fourth consonant, a fourth leg, so we know it is a foreign word. Not only does it refer to a foreigner, a translator, an outsider, it is itself foreign. It does not belong to the Semitic language, which is an elegant tool. In a Semitic language words have three consonants, three corners, and can hide what is between them, the unwritten vowels that call up the meaning.

We drive on, heading southeast. The road follows a broad sand basin a mile or so across. The sand is hardpacked with colored pebbles between low sandstone hills. The Kanayis temple is way off on the south side of the road, cut into the face of a cliff. It is a water marker, built to commemorate a well dug two thousand years ago.

The caretaker is a small, soft-featured man, with light brown skin stubbled white and one long tooth in the front of his mouth. He has a trilling, cracked voice, and waves his arms as he talks, swaying out to show us the temple. He tells us he has been alone here for twenty-seven years.

The face of the cliff is dramatically wind-weathered sandstone, the texture rilled in fine lines up close. From a distance the color swirls in pale to deep yellows and reds. The wind has eaten it away from below, forming a blufflike crest. I gaze at the lovely surface pattern and fancifully make out shapes in it, phantoms at first, but the hammered outlines of their edges are definite. Suddenly I find myself looking at a procession of antelope. Two rocks joined into a sharp corner are covered with dozens of horned animals. High up in the cleft are crudely cut cattle from an earlier time. A Greek scene is below, cartoonlike and quite beautiful—a man playing a pipe before a fat dwarf, and beneath it a plump, small-horned gazelle.

We drive farther east into the low red hills of Wadi Baramia. A man in a mint-green gown lies by the edge of a huge columnar rock, in its strip of shade at midday. The rock lies where it fell ten thousand years ago. On its side is the drawing of a horse with a long spiky mane.

The sandstone ends and we are in a landscape of sharp granite peaks, an earlier low grey formation, and, cutting through it, red granite mottling the mountains. There are walls of black, green, and silver spangled rock against the wide, pale sky. Acacia fill the sand wadis that run between them, trunks low and twisted, stained with black.

We stop under a large thorn tree to rest and let the car cool down. The Niva's temperature gauge is high in the red. Gamal drove Russian jeeps in both the 'sixty-seven and 'seventy-four wars, and has an unswerving faith in their ability to keep going. When it boils over, he tells me, you simply turn it into the wind and make tea.

The ground beneath the tree is red with blood. "The Arabs have slaughtered a camel here today," Gamal says, fingering the small soft-colored rocks, smelling them. But his wilderness instincts are off. The blood is sweet and sticky, and we see that it is the blood of the tree. Gamal finds a swallow in a crease in the trunk and scoops the bird up with his hand. It is warm and alive, and throbbing with exhaustion.

A few miles down the road, beneath another tree, a woman with a gold nose ring sits with two children. An old man hits the top of the tree with a long staff, two whitened stripped branches stapled together. "What an exhausted country," Gamal says wearily. "Who would live out here? No video. No movies. No Egyptian woman would live out here, I can tell you that."

A road branches off toward a place called Hafhafit at the mouth of the Wadi Um Khariga. This is the road to Hassan a Shazli. We do not take it, though. Instead we drive through a rose-pink valley toward Marsa Alam on the coast, fifty kilometers away. It is evening now. The sea wind sends a pair of ravens spinning across the road. We sense the presence of the sea, smell the salt in the air, and see the reflection of the deep water on the lavender underbelly of the sky.

We sit on a rock and look out over the breaking waves—a

white froth on a thin veil of turquoise over colored stones. Rock crabs cling to the coral cliffs as the water crashes around them. Their green backs are veined with yellow. Their claws are blood-red. Hundreds of "flower fish," jellyfish with purple rings, wash ashore. We handle their soft dissolving bodies and toss them back into the water.

At the geological survey station, Mr. Rifaat, the director, gives each of us our own cottage on the sea, a pasteboard house in the sand, surrounded by barbed wire and broken glass and half-smashed stinking conch shells. The thin walls buckle and sway in the shrill whistling wind.

In the evening Mr. Rifaat sends over his cook to make us a dinner of sweet fried fish. There is no water. Only the bottled water, warm as tea, that we bring in from the car. The water in Marsa Alam comes in a boat from Gebel Massala once a week. There is never enough of it.

"Water is the most important thing," Gamal says. "Does not the Quran say, 'Man is made of water—water and mud.' "

We drive on in the morning toward Hassan a Shazli. The acacias along either side of the road are uniformly withered and brown, like the brown crumbly ground out of which they grow. The ground has not seen rain in years. The mountains on either side are tall, pyramidal, and almost metallic black in the late-morning sun. Here and there are clusters of shacks made of ridged aluminum siding nailed together with scraps of wood where families of nomads live.

At Hafhafit we leave the road behind. We drive across a field of sharp black stones and into a wadi that runs toward the escarpment. We are entering new territory, and feel a giddy sense of freedom. For Gamal, who is now in his fifties, I think it brings back the memory of driving a jeep twenty years ago in Sinai, the territory of his youth.

As we come around a bend in the scorched wadi bed, in which all plant life is withered down to the root and even the camel thorn comes up in dry yellow-brown clumps, we are astonished to see a leafy green tree growing against a light grey granite outcropping striped with white marble. We stop and walk over to see what kind of tree this could possibly be.

The air as we near the tree has a lovely smell, faintly like honeysuckle. There is a droning buzz in the air all around. The tree's creamy yellow flowers are full of black wasps and flies, as though it has drawn to itself every living thing for miles. I climb the outcropping to see if there are other such trees in the wadi, but cannot see another one like it.

When I come down, Gamal is sitting barefoot up in the tree. He has become relaxed in the desert, and expansive. He laughs delightedly and tosses to the ground what turns out to be a small bird. I know the bird. I see it sometimes in flocks in the south at this time of year: a white-throated bee-eater. Its shoulders are a malachite green and gold, its wing feathers turquoise and blue. The little bird's desiccated body is speckled with blood.

"What was the name of those birds we saw this morning?" Gamal says. We had looked up and seen a flock of large birds flying soundlessly over Wadi Um Khariga. Cranes, I thought. "There's one up here in the tree," he says.

I look up and see a fluff of fawn-colored feathers striped white and black curled into a dense ball around an upper branch. Gamal dislodges it with a dry stick, and the body falls airily and lands with a muffled thud at my feet.

It is an eagle owl, mummified, dead perhaps a month. Its beak is buried in its belly. The owl impaled itself on the long, sharp thorns that are hidden by the tree's thick covering of green leaves, and died trying to bite itself free.

"What kind of tree is it, do you think?" I say.

"Shagara tyur maitin," Gamal answers, naming it—"the tree of dead birds."

The Niva will not start. The battery is dead. Gamal by now has assumed a sort of priestly authority over the car. He picks up a piece of pink-flecked granite and thumps the engine a few times.

The engine light goes on. He flashes a smile of triumph. Tiny insects—greenish-gold, like leaf-hoppers but paler—fly up as the motor starts, and cover my left hand.

Driving on toward the escarpment, we pass what seems to be a heap of rocks. A flag on a pole is wedged in at the top, and Gamal says it must be a sheikh, the tomb of a holy person.

The ground around the tomb is littered with animal bones, jawbones and camel skulls and rams' horns. Among them are circles of stones containing the charred remnants of cookfires. There is the faint sweet smell of incense around the rocks in the dry air.

The tomb is round and built of large flat stones, broken granite slabs from the mountain Hafhafit, pink, black, and red, piled without mortar. Shreds of cloth and wads of hair and paper are stuck into the cracks between the stones.

We bend into the low entrance and are at once delighted by the coolness of the shade inside. We sit on the sand floor, leaning against the wall. We stretch out our legs and have a welcome rest. Gamal is muttering and davening beside me. He has never seen anything like this before. It is clearly, he says, a holy place.

A coffin stands in the center, beneath the slight rise of the dome. Four branch poles are set in a rectangle around it. A white sheet is stretched over the top. When our eyes adjust to the light we see the branches and the thin twisted trunks of thorn trees streaked orange with henna that are unevenly set across the tops of the walls to prop up the roof. Shreds of green cloth are tied to them, to represent the color green.

A sheikh does not need a whole tomb, Gamal explains, only a sand mound surrounded by little rocks with a big stone in the corner nearest the road, and above it a flag, "so that it is seen." The main thing is the dead body, and near it some source of life, a spring or a tree, and some noticeable thing, unusual aspect of the landscape, a mountain for example, like Hafhafit. But since there is no green thing here, no tree growing nearby, there must be green cloth.

Ragged turban cloth as well, a yellowed white, is wound in around the branches. Camel hobbles are tied to them on the left side, with hair from camel beards tucked in, markers of animals sacrificed to the saint.

Rusted cans, a blackened cookpot, tea glasses, a tin canister of sugar and one of tea are stuck up in the rags. Plastic oil containers filled with water stand in one corner. Beside them is a glass jar in which there is still a mix of twigs and dried leaves and frankincense, looking very much like the quartz pebbles scattered on the ground outside.

The torn backs of cigarette cartons are tied to small branches around the coffin, set up as votive tablets. Their plain sides are covered with names and prayers, and here and there the picture of a large animal with long curving horns, resembling the rock drawings of ibex we have seen on our way here. Gamal writes our names on the cardboard beneath the others.

We hear a noise outside. Someone crouches to enter the tomb. It is an Arab, as Gamal would say, a local person. After the formal Islamic greeting, the man introduces himself as the nagib, the caretaker, of the tomb of Sidi Salem, which is out by the main road. He clearly wants an explanation of what we are doing here.

"My sister Suzanne here," Gamal begins inventively, and I wince, knowing he is about to indulge himself in some fabulous yarn at my expense, "is a Muslim from America. Sidi Hassan a Shazli appeared to her in a dream there and she has come all this way on a pilgrimage to his tomb. I am an Hagagi from the city of Luxor, and have offered to accompany her for her safety. Can you tell us, please, kind sir, who is buried in this tomb?"

The caretaker seems not only satisfied with the story, but impressed and newly deferential. He tells us Ababda in a camel caravan found this man lying in the desert near Hafhafit covered with dust and thought he was asleep. When they reached him they saw that he was dead. He had been dead for five days but looked as if he were alive or had died only a moment before. They put him on the back of a camel and went on. Where the camel stopped they buried him.

The nagib says that the Ababda come on Thursdays and Fridays for feasts and prayers, more often in winter, when the tribes are on the move. The tomb is called Al Tum, "the twins," though he does not know why.

Gamal, in his new expansiveness—an intellectual of folklore, after all—regales us with the Nile Valley story about twins. "When twins are born, their souls slip out into tailless cats at night. If someone hits the cats, the twins feel the pain. If someone locks up the cats, the children don't wake up. The mother runs around the neighborhood yelling, 'Let out all the cats,' and everyone knows the souls of the children might be in them."

We offer to drive the nagib to Sidi Salem on the way back to the geological survey station. We will not make Hassan a Shazli tonight.

The Niva boils over in a long stretch of sand. Gamal stands over the motor, cigarette dangling from his lip as he mercilessly feeds our precious drinking water to the engine. "With the baraka of Sidi Salem, and the baraka of Sidi Hassan a Shazli," he chants, "and the baraka of Saidna Musa and Sidi Al Tum and Saidatna Um Ghanam . . . we will be moving shortly."

For half an hour we stand silently, chain-smoking in the hot sun. When I try the ignition again, the jeep miraculously starts.

We reach the tarmac road by late afternoon. Our spirits are

high as we roll down toward the tomb of Sidi Salem and the mouth of the Wadi Um Khariga. Tribesmen are playing dominoes in the shade beside the tomb.

We do not notice the young man in khakis sitting a little apart as we climb out to join them for an Orange Fanta.

The boy is a soldier. He wants a ride, a hundred or so kilometers back over the desert to the border outpost of the Mukhabarat, the secret police, at Hassan a Shazli. Gamal is suddenly obliged to abandon his air of authority and revert to his self-abasement before armed children. Gamal gives the boy soldier cigarettes, buys him a Fanta, eventually tells him he doesn't think the jeep can make it. It just boiled over. We don't have enough gasoline. Do we have a permit for this area, the boy asks, then tells us we are under arrest.

I am still playing the high-handed foreigner. I laugh at the soldier in disbelief. I get up and walk away. The boy springs after me and, twisting my arm, wrenches the car keys out of my hand.

The soldier forces us to sit, for hours, at the tomb. He struts around us. He lists our crimes, forming his story by repeating them over and over, improvising, persuading himself of our misdeeds. Where was our tasry, our government permission? No one is free to wander in the desert alone. We are in serious trouble. And so on, an interminable harangue.

At last, when the boy has exhausted his imagination, Gamal takes him aside and offers him a pack of Cleopatra cigarettes and five pounds, saying, "We are all human. We all have to eat."

The boy exploits the moment. He pretends to accept the cigarettes and money. Then, with a flourish, he throws them into Gamal's face: "A bribe! You are trying to bribe me?"

Gamal for the first time is genuinely frightened. His carefree pleasure in our adventure has vanished. He is right to be anxious. The Mukhabarat can destroy him, if they wish—close his shop, revoke his shopkeeper's license, lock him up for a few years.

Near midnight we drive off in the dark down the road to Hassan a Shazli—I at the wheel, Gamal in the passenger seat, our captor, Ali, in the back. The car lights reflect green eyes in the dark. I make out the soft round outline of a small gazelle.

We arrive at the grey concrete bunker of the Mukhabarat at Hassan a Shazli. Ali disappears inside.

Presently, two older men emerge. One is a paunchy, cheerful-faced Nubian, the other a fair-skinned Alexandrian, with a blue dent in his forehead, the mark of piety. Both men are soft-voiced and kind. They ask a soldier to bring us tea and bread. The boy soldiers train their guns on our faces.

Gamal repeats pleadingly, over and over, "I didn't try to bribe him. The five pounds fell out of my pocket when I took out my cigarettes." He embarks upon a long, impassioned, elaborately untruthful version of our trip: of how we never left the main road, never went to Hafhafit. I see that he is vigorously unraveling the narrative fabric that young Ali went to so much trouble to weave.

The two officers listen impassively. They calmly tell Gamal to sit down and relax. What is he so afraid of?

I have a hacking cough. But it subsides. I drift off to sleep on a bench.

I wake a little later, still hearing Gamal's voice. He is explicating the Quran to the soldiers who stand guard over him. He davens slightly, and breaks into chanting as he talks. The Quran is the one thing he has to fall back on. I feel a sharp stab of guilt: I have selfishly led Gamal into the worst nightmare he can imagine.

We have not seen each other's faces for hours in the dark. A man in thin white pajamas comes and sits with us, his knees in desert bent posture. He has brought us tea. I talk to the soldiers about desert animals—scorpions (there are many), the drayshas (vipers)—they are blind and move in waves and live in loose sand. Wild dogs—they roam the desert and eat anything they can bring down. They eat Arabs they find walking alone at night.

A rooster is crowing. Dawn wafts softly in, lavender over the mountains. Pigeons veer in even circles down to the road. Across from us the Ababda village unfolds to the morning. Donkeys whine. A string of white camels against a distant pink mountain, moving, moving, stop to tear at the soft upper branches of acacia trees. An Ababda woman wrapped in emerald silk squats over the dawn fire to boil milk.

I am caught up in my own thoughts of freedom, the bright, glowing moments of my life. The lake. The marriage months before on its shore. I imagine, as I often do to orient myself, the deep strip of blue between hills of woods and fields and old red barns—north to the town, south where Scorpio rises in the summer, west where the sun sets, east behind where I am standing on the shore. The smell of cold and weeds and lake water snaps into my head, detaching me from where I am, and for the first time in this experience of being held captive, I am calm.

Sitting still, I have watched the sun set, the wind rise, the moon rise, the procession of familiar stars, the evening birds come and go, the night birds follow them, and at last the stir of the morning—the wind of dawn, the birds of dawn. I have been forced to sit still. My frustration and fear now melt away. In the night I decided to leave, to run when I was set free. To go to Cairo at once and leave the country.

The first streak of sun lights up our faces. My white shirt and blue jeans are streaked with soot, my hands and fingernails encrusted with dirt. Gamal's sweet anxious face is creased and covered with white stubble. The man beside us is Said Abu Magim al Rifayia. He has a rubbery, hideously ugly face, covered with deep wrinkles. One eye is slightly off. He tells us he is twenty-seven years old. There are healing sores all over his body. He whispers to Gamal not to worry, everything is going to be all right.

Men begin to get up, to wash. I take my Egyptian dictionary out of the Niva and tear a blank page out of it and begin to draw. I draw a nearby acacia, the sap-streaked trunk, and then, turn it into a pear tree.

The soldier standing over me sticks his gun in my face. "What are you drawing?"

"A kind of tree that grows in America."

"I want you to draw another one right here."

He points to the page. I turn toward Gamal, ignoring the soldier, who is torn between his impulse to bully and his uneasiness at the sight of the foreign writing beside the tree. I have written, "I have fallen in love with American names, the sharp names that never get fat, the snakeskin titles of mining claims. . . ."

In the background the commander's voice drones on the radio to his superior. The line has finally gone through to the main outpost at Berenice. The conversation is abruptly over.

The officers emerge, utterly changed now. They fawn on us, they put cigarettes in our mouths and shake our hands. They have been ordered to release us immediately.

We scramble into the Niva, shoeless as we are, as fast as we can. "Let's put Ali under the car," Said Arifayia says. We drive before the soldiers can change their minds.

We make our way west with the wind behind us. Gamal, with tears in his eyes, says fervently, "A whole night at Hassan a Shazli! It is true what they say. The desert is a place of miracles."

He rides for a time in silence, staring out the window. Then he says quietly, "You know, they really were a group of very nice boys. God bless them."

I left Luxor a few days later. I went to see Gamal at his shop. When I saw his bony back sweeping dust and stones into the street, I felt a sharp stab of love.

WHEN I GOT BACK TO CAIRO, I met a man named Joe Hobbs. I had been trying to get in touch with him for months, leaving messages for him on his answering machine at the University of Missouri, where he was now a professor of geography. I had never met Joe, but I had been hearing about him for a long time. There was the story of his famous bird dinner in the early eighties: He went to the market in Alexandria and bought winter migrants—pelicans, flamingos, herons, the unclean birds of Deuteronomy—and served them up at a dinner party for his American and Egyptian friends. Of course they didn't know what they were eating; he wanted to see just what people would eat, what a food taboo was all about.

"Well, the skins and feathers went to the Field Museum in Chicago," he told me the night we met, in the Omam Moroccan restaurant, drinking bitter hot chocolate out of silver cups. "Nothing was wasted."

That was Joe. Scholar on the one hand, deadhead on the other, someone who had the American road tradition in his bones, the tradition of humor and simplicity. He already had the desert virtues, and when he came to Egypt he knew he was home.

Joe reminded me of a man I once met in a tea shack by the tracks at Abu Hamed, a Jaaliin merchant named Omar. I jokingly told Omar as we sat among the goat women in the shade of the nim trees that I was going to Meroe. He, jokingly, responded that that was a long way. Then he made sure that I got there, accompanying me across the desert without being asked, seeing that I was fed and protected, even sleeping across the threshold of the doorway of a room I took for the night somewhere while I slept inside. He graciously saw me to the outskirts of the city that was my destination. Then he turned around and was gone.

Joe had the same desert instincts. He was a companion who would, without a second thought and for nothing in return, go hundreds of miles out of his way to show you the road. He knew the Eastern Desert, and I had hoped he might be able to give me a contact among the Ababda there.

With his thick growth of beard and thick gold-rimmed glasses, and face and hands leathered over by the sun, Joe could well be any one of the wanderer naturalists who have walked through the remote mountain and desert regions of the earth for the last two hundred years. He was just back from six months in Sinai, where he found the only other known specimen of the "burning bush," *Rubus sanctus,* a desert-growing raspberry, some miles seaward from the Monastery of St. Catherine, where the original is preserved.

That is why I could not find him. He had been in Sinai all this time. My first night back on *Fustat* after the misadventure with Gamal, the phone rang at midnight and it was Joe.

Over dinner he told me how he had first become interested in the Eastern Desert. In 1982 he traveled with a Dutch ornithologist who wanted to find out where the white storks landed along the Red Sea on their yearly migration between the steppes of Central Asia and East Africa.

Maaza tribesmen, the only people who knew the answer, took Joe and the Dutchman to the Wadi Malah, the salt flats northwest of Hurghada where the birds come down before cutting back over to the Nile Valley and south over Aswan and Lake Nasser.

Joe understood then what his work was, "where I wanted to be, and who I wanted to be with." He asked one of the Maaza, Saalih, if he would come "work with him"—to walk with him throughout the region. They began a series of thirty-to-forty-day journeys over the years exploring the whole Maaza territory, the Eastern Desert of Egypt between the Nile Valley and the Red Sea north of Quseir.

In the Eastern Desert there are remnant stands of trees. "Stranded trees," Joe called them, telling me of an ancient wild olive on Gebel Shayib, four giant ziziphus (crown of Christ) trees in Bir Um Sidr, planted by the Romans two thousand years ago. What interested him was that the Maaza had not cut them down. Charcoaling is one of the few ways Bedouin can make money. The trees were, in a sense, he said, sacred to them.

Saalih led Joe across the south Galala Plateau to show him a cave where three Romans had been turned to stone. Inside, Joe found huge stalagmites, curved and humanlike, glowing white in the dark. In a cavity behind them were the fur, bones, and teeth of twenty-seven leopards preserved by the salt and heat. Months later, in a lab in Chicago, they were carbon-dated back ten thousand years to the Neolithic.

The Galala is limestone, in places white and crumbling as though streaked with snow. It is pocked with caves. Galala, galena, the name suggests shining. In Greek, hence Coptic, gal=kel=cell—the place of cells. The anchorites came here. St. Anthony's and St. Paul's, the earliest Christian monasteries, formed around caves on either side of the plateau.

Below the Galala on the north side is the Wadi Arabi, where fallen coils of rusted barbed wire mark that the land was mined during the war with Israel. Every year a Peugeot station wagon of tourists or a Maaza camel is blown up in the Wadi Arabi.

As a teenager Joe thought about going into Egyptology. What interested him were the marsh scenes in the tomb of Mereruka at Saqqara. He wrote his undergraduate thesis at Santa Cruz on the animals represented there, and then came to Egypt to see what had survived. He struggled with questions like, "Are there, were there ever, wolves in Egypt?" and then searched through the great catacombs of animal mummies north of Minya to find out if any had been buried there.

He told me about the lists of Maaza names that he had collected on his walks over the years: the hidhid (the eagle owl), the hidhif (Burton's carpet viper; "Looks like a piece of red granite. You're dead in five minutes"). He learned the best way to travel in the desert was to hear about something and then be led to it, to the wild fig in the high mountain glen or the scatter of scrub that survived for a year after rainfall. Often the tree or the scrub had given its name to the mountain or the wadi bed.

Like me, Joe had married right before he left on a grant. He had been looking forward to going home to his new marriage for months. Now he was back in Cairo. His fieldwork was over. He was beginning to feel the familiar dread that came with knowing that it would be a long time before he would be able to get back to Egypt again. Egypt was a difficult place, but for us it was a place of tremendous freedom. He had been living a rough life outside for months. How could he go back to an office, to a city in central Missouri?

When I asked him whether he had any friends among the Ababda he said he did not, because they came from the south. But he knew of one family. He told me this story:

One winter he and Saalih went to climb Shayib el Banat, the highest mountain in the Eastern Desert, a mountain so high that sometimes in winter its peak is covered with snow. On the top they found a package wrapped up and placed under a stone. Inside the package was a crystal vase, and a letter from a European mountaineer Pierre Rosenthal. Rosenthal had left the vase in 1966 as a gift for the next person to reach the top of the mountain.

The package had been placed beside the old cairn. Joe and Saalih decided to rebuild the little heap of stones. As they took the cairn apart, stone by stone, they found another letter. This one from George Murray, the Scottish cartographer who had mapped the Egyptian deserts in the 1930s.

Joe found out that Murray's wife, Edith, was still alive in England. He called and arranged to visit her on his way home to the States that year. She was dead before he arrived. He was informed that she had left a sum of money to a family of nomads in the Eastern Desert, the descendants of Murray's Ababda guide, Ali Kheir.

Joe returned to Egypt and spent two years asking after them. But he was too far north. One day Saalih heard that Ali Kheir's grandson Saad was in Hurghada for the night. He met him and arranged for his family to receive the money.

Joe thought the Ali Kheirs would help us, if we could find them.

"So let's go," he said. He had only a week left in Egypt, and he didn't want to spend it in Cairo. We would meet in Hurghada, on the coast, in two days.

I arrive at St. Paul's just as the sky and the sea and the land are all melting together into a creamy blue, a fuzzy delicious blue light. I arrive just as the Niva runs out of gas and rolls down the paved wadi bed that cuts through the shaggy cliffs surrounding the monastery. The Niva sputters and dies in the bowl at the feet of the old mudbrick towers. The towers look, after thirteen hundred years, like something made by a child and stuck unevenly up against the Galala Plateau. Its crumbling white sweeps of rock crouch down like the forepaws of a gigantic animal around them. A monk comes out of the monastery door. He has a black beard and a black bonnet and a red jerry can full of gasoline.

It is exactly a day's drive for me in my slow-going car from

Cairo to St. Paul's. I often spend the night at the monastery when I am driving south. I walk up the stairs of the annex to the room where I have often slept. The key is in the door. I light the row of half-melted candles I left stuck to the window frame months ago. The sea wind buffets the little flames so that their light dances on the high wood beams of the ceiling. A spiny mouse squeaks across the floor. There are ten cots in the long room. As usual, I make up an elaborate bed for myself with my sleeping bag and the pillows I have brought. I go down the stairs and wash in a bucket of cold water before going in to supper.

The cold November wind blows across the dark courtyard of the monastery. The wind stirs a feeling of deep loneliness. I know it so well. And tonight I love it, for I sense that it means I am going off into a strange new territory.

In the refectory I am given a plate of soft white cheese and coarse brown bread and honey. On the wall is a picture of St. Paul in a palm-leaf garment with a lion on each side, for the old hermit was, in the story, buried by leopards. Father Agathon tells me tonight that the Galala is known locally as Gebel Nimra, Leopard Mountain, and I think of Joe's story about the cave. I go quickly up to bed. I will rise with the bells at three for the dawn prayers, and the drive to Hurghada, four hundred kilometers to the south.

I drive down the one long street that runs into the seaside town,

past the central mosque with its stripes of green neon. I see Joe on the street. "Let's get out of Dodge," he says, as he climbs in beside me. We stop at a stall and buy some things, tuna fish, a metal teapot. We turn the car around and head for the Qena–Safaga road and Mons Claudianus, where we will camp tonight. He has heard from some people in Hurghada that Saad is in Wadi Kharit, near Kom Ombo.

In the days that follow, as we travel together I remember why I have longed for the desert, its immense stillness, the sweet radiance of morning and evening, at night the rich darkness around me as I lie on the stony ground beneath the thick stars.

Joe has been telling me about Elba, the mist oasis. Elba is a desert mountain near the sea. It is so often shrouded in mist that plants grow there that grow nowhere else in the world, plants that come out of the gravely dry ground and thrive on the moisture in the air. The animals and birds drawn to them come from the south. Hence the mountain is an island of Ethiopic, African, fauna in the middle of the eastern Sahara.

The Elba massif is the central feature in the tribal territory between Egypt and Sudan, and difficult to reach. No road goes there. There is a legend that the mountain is a place of origins, is itself a living being, the ancestor of the Beja. A moss-covered cave on an upper slope appears to breathe, to shoot forth sprays of steam as though (some have guessed) there is a geyser inside. The Besharin who live on the mountain will not let anyone near the cave.

One afternoon in November in Wadi Kharit, Saad agreed to take me to Elba. He reached into the pocket of his thin white gown and brought out an ostrich-feather fan, the soft inner grey feathers of an ostrich tied together with dime-store string, and handed it to me.

He suggested we go first to the sea and his cousin Saad Abdullah, who lived with his mother at Abu Ghusun. Saad and his cousin were Ababda of the Nefaa tribe. Their home territory was around Gebel Hamata, a mountain north of Elba on the coast, but they had relatives at Elba. Saad laughed when I told him why I wanted to go there. Elba was unusual, it was true, but he would show me all the green places along the coast, the mountain glens and what pockets of water were left from the last rain, some years before. His relatives around Hamata would walk me through the mountains, and he would come and meet me there.

I would arrange with the Cairo Herbarium to make a collection of plants from the area for them. This would give me, after long negotiation, the permits I needed, and a way to track the path of water, hence plants, animals, and people. In the Red Sea Hills every living thing follows the rain. Desert plants have differing abilities to retain moisture and last accordingly, so that the desert is a mosaic of dead and living plants. Elba was covered with living plants, was a green mountain in a sandy desert, a paradise somewhere south on the Red Sea, veiled in mist.

Joe and I drove back north the long way, sweeping through the oases of the Western Desert, stopping for a night at each, Dakhleh with its hot sulfur springs, Farafra with its limestone cakes that look like melting heaps of snow. Joe told me that when he was a boy in Saudi Arabia there was a path of red sand that filtered down through the mountains, a magical road of sand.

The Niva breaks down in the middle of the road outside Bahamia. The battery is dead. After a while a lone trucker comes along. He unwinds his headscarf and ties one end to our fender and the other to his truck and drives off, intending to tow us to Cairo. We have no lights, no horn, only a stiff steering wheel. Joe uses all of his strength and concentration to control the wheel and keep us on the road. The headscarf soon rips and we are left behind. Perhaps we will spend another night out here after all. We decide to push the Niva off the road, into the sand. As we are about to do so, the trucker, who has realized he has lost us, reappears. We proceed thus: towed, lost, recovered, again and again through the night. Until the final break at four in the morning, beside the pyramids, on the outskirts of Cairo. Joe gets into a taxi. He has to attend a conference on the geology of Sinai in Ismailia later in the morning. He is gone.

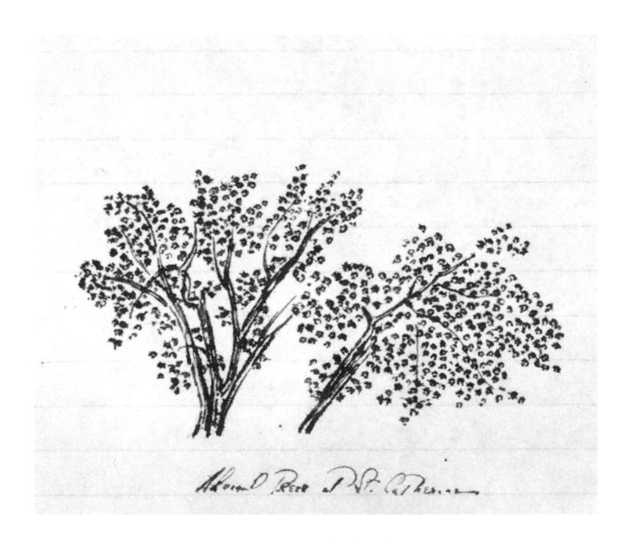

GIFTS FROM THE SEA

THEY HAVE SKIN LIKE BENI ADAM but it is green, and they breathe like Beni Adam. Their eyes are gold. They have fingers and hands. They live in sweet water. They come out after rain and they sing all night long."

Abdullah takes me through the wadis around Hamata, the highest mountain in this part of the Eastern Desert. He shows me wherever there is water and the life that it brings, the pockets of life. The hawthorn trees in the stony bed of Shartut, where eagles circle silently above. Um Disi, the mother of cattails.

In the afternoon we sit under a tree and make coffee. We already have at Hamata "our samra" for this purpose. It is beside a high slab of pink rock, a small twisted tree with thorns like tiny fishhooks that catch in my hair as we bend beneath it.

Coffee takes up the hot part of the afternoon while Abdullah tells me about the plants we have gathered, the creatures we have seen, things that have happened in the past, and about the tribes: the Nefawiin, like Saad and his cousin Saad Abdullah; the Greyjab (who are fishermen and famous for magic), the Kreyjab, the tribe to which Abdullah belongs.

These are the tribes of Hamata. They have always been together, as far as he knows.

He tells me that in this ravine at Shartut he once came with twelve other men to track a wolf, a very large wolf that was eating their animals. They could hear it howling at night and followed the sound.

I take out one of the pictures I always carry with me, of my old friend Gary Lynch holding up an Eastern coyote he has trapped behind our cottage on East Lake Road.

Yes, Abdullah says, that is the animal, but bigger than that. Its color is the same, agabash. A dog will eat out the heart and liver of an animal, but a wolf takes it by its throat, drags it away, and eats the whole thing.

Abdullah's skin is the dark, dark skin of the mountain people that one finds moving south and inland, a matted black, with no shine. His face is very fine, with a small beaked nose, a falcon nose. He carries an antique pair of binoculars hidden in a pocket of his yellowed cotton gown.

I have read that there are people who fish, and those who

pasture animals, and there is no mixing up of the two. The Greyjab are fishermen. But here even in one family there are those who khabar elbahr, who know the sea, and those who khabar elgebel, who know the mountains.

I knew two families of fishermen. Ali Hassan, Saad's cousin, and his mother and brothers. His mother was the one who taught them how to fish. The other was the "harim" at Abdullah's: Halimi, Selimi, and Shaya. Abdullah himself would not set foot in the sea. He said that the sea was the province of women.

Abdullah's daughter Shaya was the fiercest of the fishermen. She was very fierce-looking, with a terrific, bitter sense of humor. She wore at home an orange silk sari from Shallatein, but in the sea she wore a black dress, soaked through (though this did not seem to slow her in the water). She carried her net heaped over one shoulder, and a long rusted metal pole which she stabbed through the eyes of the fish she caught, using it like a stringer. She had a long sharp knife bound around her waist to pry open clams (gurmat), and to chase away the sharks that got too near.

The fish Shaya usually went out for were the green unicorn fish, called here Abu Grun (the father of the horn—as they have a long hornlike back fin). They are, I think, a species of parrotfish; they have buckteeth—with which, David once told me, they eat coral.

Shaya would go out when the water had receded some and was level with the seaward edge of the reef. The fins of unicorn fish would appear along the reef wall, a large school of them. We would spread the nets, and Shaya would rush ahead, barking and hitting the water with her metal wand, and drive the fish into them, and then we would haul them in.

The illustrated sea. Standing on a white cliff we watch the giant colored fish rise and fall in the water below, their color wavering, spreading like shimmering cloth stretched out beneath the surface: mammoth rays like shadows, Um el Ghattas, mother of the deep, Abdel Hadi tells me they are called.

Sometimes we come upon a boxfish, a sandu, which is very fast in the water. It looks as though it was deliberately made, made by hand as a lovely joke, a gift for a friend: apricot-yellow with white-rimmed blue spots, and shaped precisely like a box, its huge eyes popped out, its fins translucent and small.

Sometimes we would find the drumpa, a blowfish. Shaya would pick the poor puffing fish up, simply pluck it out of the water with her hands, and clean it on the spot with her hunting knife. She holds up its dripping liver to show me—"fee sem," there is poison in it—before tossing it back in the sea, running the flesh through the water and dropping it into the plastic burlap sack I am carrying.

We only watch out for the moray eel, the shayga, the draysha of the sea. And the Ugam which comes in very fast and bites your legs off, bites off the arm of the fisherman that tries to pull it in, and is gigantic and an almost translucent blue. I could never figure out what it was.

Poison disguised as beauty. There were many stunningly beautiful fish, a warm striped orange with a black back fin—so diverse I cannot remember or describe them, all seeming carefully painted, fancifully thought out, crafted by hand. Many had poison in their fins, some hidden source of poison.

"Never go down to the sea without someone who knows the sea," said Shaya.

Sometimes we would make a charcoal pit on the shore and bake a fish under the sand for lunch. Tearing its sweet flesh away from the delicate bones with our fingers. This was at Marsa Rodamrai.

Beyond the bay is Gebel Hunkorab, purple-grey in the light of noon, with loose cream sand spilling over its peaks.

As the beach curves south it is striped with plastic: blue, black, white. The northern curve is clean. The wind comes up as the waves come in. The waves bring the wind says Shaya.

There is the camel of the sea, it looks exactly like a camel. These can be eaten.

There is also the Arussat al Baher, the bride of the sea. We cannot eat the flesh of Abu Salama or the Arussat al Baher, because they are in form like a human being, like Beni Adam. They do not eat fish but bersim (marwa, grass) of the sea. You see them, all the large animals, on a calm day. They breathe air, like the bissa. It also breathes air and stays out of the water for forty days. It lays its eggs in the sand. Next month this happens. The little ones crawl to the sea like the bishbish (the hermit crab in the snail shell). They grow to be enormous. You can catch them from the side and flip them over, but not under the upper (right) arm, or they clamp their arm down on you and drag you into the water to drown. You open them up just like a box. On top of them and underneath is hagar, rock. They must come up and breathe, that's how you catch them. Imagine, they breathe as we do, and yet they live in the ocean all the time.

Shaya pries the shell off a dead sea turtle that has drifted in, but it is too scratched to sell or use. In the house there is a turtle skull, and a tiny baby girl with cowrie shells in bracelets on her wrists and ankles.

Sometimes we go back to the house, to Abdullah's camp in a wadi just up from the shore, protected by low dunes from the sea winds, and hidden by them. The camp is a circle of shacks made from scraps of wood that have drifted up from the sea, nailed together in the shape of small houses (as though the material did not matter, only the form of the house). The wind pours insistently through every chink and crack. But there is shade and coolness, and nails to hang things on.

Inside Abdullah has painted the walls and objects with blues and greens and yellows, striping and mottling them. The house has an air of cleaning and cooking, of women well cared for. Swaths of bright synthetic silk hang here and there. A tabby cat. A small black goat. A pigeon tower outside, a wooden box set up on three stripped trunks.

Shaya shows me a rug she is weaving on a stick loom. The colored thread is spun from old worn-out dresses, making orange and red stripes.

Halimi is braiding up Selimi's hair, in tiny "pharaonic" braids, matting the hair with crumbled charred camel dung applied with her wet fingers.

A kaf miriam, a hand of Mary, hangs from a leather cord in the middle of the room. Halimi tells me that when it is green and flowering it is very beautiful, and as it dries and its leaves die it closes, slowly, like a hand.

These are Abdullah's summer quarters, by the sea. In the winter they pasture their animals up in the desert foq. I have already seen, walking with Abdullah, some of the stone circles of years past, walking by them as Abdullah remembered the years. The stones where Medini was married. In a wadi of greenish gold we pass a tree in which there is hung a crook and some cloth.

"I left them there two winters ago," Abdullah says.

We can often find the remnants of their camps by the white crusting of vulture scat over the nearby trees. Vultures follow human beings, and eat what they leave behind.

I met Abdullah and Halimi in the winter in Wadi Hagaleig, the wadi of the two hagleig trees.

Medun was then recovering from an illness, confined to a kheima, a low dome of bent acacia branches covered with rugs. Halimi led me inside, out of the bright noon sun, and I was amazed to find it lined within with beautifully colored cloths, and to see there Medini, their daughter, a beautiful girl with honey skin and long dark hair, reclining on pillows and playing with a baby rabbit.

THIS WAS THE FIRST TIME I met Abdullah, Halimi, and Medun.

Saad Abdullah brought me to Wadi Hagaleig to ask Abdullah to take me to Sikeit, at the Emerald Mountain, where the old emerald mines were and there is a strange temple to a horned goddess cut into the mountain's wall of silver rock.

Abdullah was not at the camp then, and we drove off over the sand to where he was with a baby camel, one day old, and its mother, his naga and her hashi. What a soft, new thing it was, with a lamb's-wool hump and long soft white legs folded beneath it I first took to be cloth, or a thick froth of milk. The heal beneath its breast was already formed.

The mangroves are called Um Sulifat, the mother of turtles.

By the mangroves here there is a house torn apart on the

shore, shreds of wood and cloth, and beside them a small tin teapot painted blue, lying on its side in the sand.

The mangroves are huge, ancient trees. I walk under one into its network of silvery white trunks, soft as fur sprouting rust and green leaves (my bare feet loving the thick ridges of wet mud and sand beneath their skin). I am not alone, under them is a tiny hen with lovely feathers—brown, red, blue-black, a bright orange beak, stalking the blue crabs whose backs like thimble caps pop up here and there.

There is a sudden sense of danger near, not even recognizing the sense, but letting it come sidelong in images, in abrupt memories of stories about quicksand, before I consider the possibility of there being quicksand in this place, bura, the pits of loose sand. Abdullah later tells me that camels, wading out to eat the mangrove leaves, will sometimes go down in them.

Abdullah and I sit on top of a crag on Gebel Makbia. A heat haze smokes over the bowl of surrounding mountains. Makbia, Abdullah explains to me, is the dividing line between the two seas. On one side of the mountain the rain that falls, in flash floods, flows down the sharp short wadis to the Red Sea, which is the salt sea, the bitter sea. On the other side the rain flows through the long winding wadis that are the dry beds of ancient rivers running down into the bahr helu, the sweet sea, the Nile. He points to the wadi one travels to Hassan a Shazli, to the west. He says to me—this is our common joke—"Lazim teshuf taetal da," You must see that ibex. We can hear them scrambling on the rocks below, but cannot make them out.

Abdullah's territory is between the sweet and the bitter seas. He has no love for the reef, the black farmland of the Nile Valley, which is dirty and chaotic. The desert, like anything beautiful or well done, is nadif, clean.

A baanib, a lappet-faced vulture, circles below. "Byakul al baheim," Abdullah immediately says, It eats our flocks.

Drawing horns in the sand by the fire, the oryx, the ariel (he sings me a song about a man who is as strong and steadfast and wild as the ariel, who disappears rapidly into the desert), the taetal, the ibex, the wild sheep that still live in the Wadi Allagi.

He tells me there are butterflies that arise from the dew. "Fee waqt el mattar," in the time of rain, they rise in clouds from the buwal—there are so many.

We meet a beautiful man with a fine face of ebony like carved wood, with a wild look but kind. His name is Nassur. The girls knew who he was when I described the place. A strong smell of wood smoke came off him, like cedar. I said, "What is that lovely smell? Is it from a tree or from a fire?" "From both." He laughed and showed me it was smoke from charcoal he was making from a fallen seyal tree. The smell was on his skin and in the air. Later, when I mentioned this to Medun, she said, "Oh yes, you can smell it miles away—such a lovely smell."

We took Nassur with us and looked for his three camels. In Wadi Hamata was the first selim we had seen, manawer, as Abdullah says, all in flower. Or, as we often say to each other, "lit up." Inta manawer, You are radiant, full of light. Allah yanawer alleik—May God fill you with light.

There are turquoise copper veins in the chips of hamur we see as we climb down the ibex path among the selim bushes dusted over with yellow flowers.

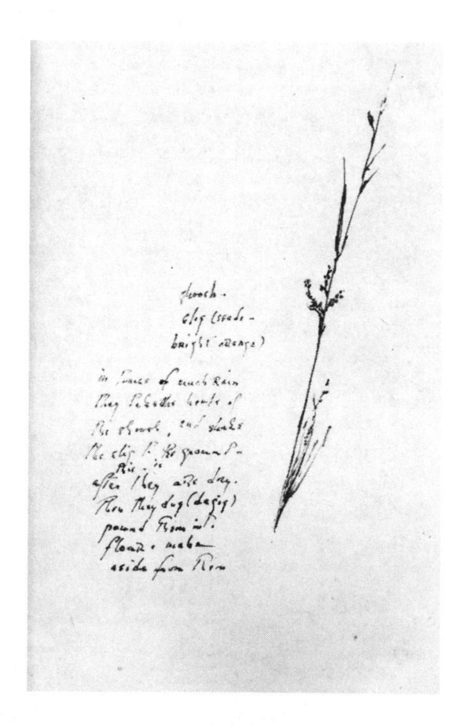

Sitting on a salt flat that looks like snow, surrounded by the live presences of trees the waves, the wind, salt on my skin. My hand is covered with henna. It looks as if it has been dipped in blood.

My maps, the maps from the geological survey office in Cairo, maps made by Murray in the thirties, flake away in my hands. Abdullah reconstructs our days remembering the wadis—taking me through them as he talks—by their color, and the plants we saw in them. He teaches me the ancient original taxonomy, that lies in the underlying qualities of things, in the similarities between disparate things: between the wing and the hand, and the shell and the ear.

ONE WINDY DAY on the Red Sea, during Ramadan, Saad and I arrive at Abu Ghusun bringing apricot paste, red meat, and vegetables from the Nile Valley. We arrive at two in the afternoon, when everyone is faint from hunger in the heat.

Saad marches into his cousin Saad Abdullah's kheima, his shack on the shore. Flies buzz up in clouds around us. The wind rattles through the loosely nailed walls of wood scraps and sheets of corrugated tin.

"We're starving," Saad says.

An aluminum tray of fresh bread, the kind of bread that is made in thin sheets over a charcoal fire, and soft white cheese appears before us. Saad Abdullah sits straight-backed, with quiet disinterest a little apart from us, not touching the food (and under these conditions: his younger cousin shows up with an outsider, a kafira no less, whom he has never met or heard of, in the hottest part of the day during Ramadan, demanding lunch). His large, muscular body is draped in fine white gauze, his gauze turban immaculate. His face, with its peculiar, slightly Chinese slant, several shades darker than his cousin's, is utterly polite. He has nothing intrusive to say. And he is clearly delighted to see Saad, who is innately graceful and offends no one. A stone from his hand is an apple.

Saad Abdullah's mother, Amna, in loose folds of indigo silk, settles down beside us when the tray has been taken away, proffering two glasses of shay bi leben, tea cooked with milk, milk that has come foaming, fresh from a sheep outside.

Saad explains that we are on our way to Elba, and we will be in the area for a while. We will need provisions. And we will need Saad Abdullah, who has water and gasoline and food.

The sun is setting bald and pale, like a planet behind me. A skate is lying in the sand offshore in the shallow water. I walk up beside it. It turns to flee. It is rimmed with seaweed yellow—its wings speckled with Copenhagen-blue polka-dots. The same blue striping on the sides of its tail—a beautiful painted toy, deadly poisonous, I have heard. It swims back from the reef and flutters in the water beside me. We move together on either side of the sand lip of the sea, watching each other carefully, and yet companionable, for a while.

At an angle, out of the path of the setting sun, I see an iridule, what my aunt Dorothy once described to me as a sun dog. You see them in late February, she said. Here, in late February, half the world away, is a sun dog, a perfect oval, a small circular rainbow. Saad Abdullah sees it too. We meet in the salt scrub between the encampment and the sea and look together at the iridule. When you see such a thing (a shamay'), he says, you must make a karama, an act of generosity, at once.

"What if you don't?" I ask.

"God is generous. I don't know what he would do."

When the last trace of the sun has gone, the fast is broken. First with dates and lemonade, things soaking with wetness. Everyone hangs back, hesitant to be the first to show the weakness of desire or need. The lemonade, made from precious lemons brought from the Nile Valley and saved for this purpose, is passed from hand to hand in a large aluminum bowl. Everyone tries to take as little as possible. It is delicious.

Then there are the usual things: tomatoes and lentils and cheese and bread.

When all this has been cleared away, the long process of making gebany begins. Coffee comes from this coast, and the making of it is one of the rituals of the day. First a small fire is made, of dry roots or charcoal or whatever is on hand.

When the fire has burned down to crumbling coals, a tuna fish or bully beef can with a bent wire handle is taken out of someone's afesh, or coffee kit. A handful of green coffee beans—grown in Yemen, smuggled up through Sudan—is removed from a plastic bag in the kit and tossed into the can. The beans are blacked over the fire, like popcorn. This takes some skill of the wrist, to black them evenly, but not burn them, tossing them lightly up and down for quite some time, talking casually all the while, as though one is quite unaware of what one is doing (for this is a social activity, after all, a way to draw people together). The blacked beans are turned from the can into a hun, a mortar carved by hand (usually by its owner) from the red heartwood of a samra tree, a variety of acacia that grows in from the sea. The carving on the hun is elaborate, with hagib, incised lines of pyramids, as a protection against the evil eye. A shred of ginger, or some black peppercorns—something hot, also removed from one's afesh—is put with the beans in the hun, where they are crushed, rhythmically, with a daggag, a pestle which is the end of one's asaya, one's walking stick, or with a long carved stone. Talc, a soft greenish-white stone found in the mountains here, is often used for this. The crushed coffee and ginger is funneled from the hun through one's hand into the mouth of a gebany pot. (I remember seeing a statue of a gebany pot, with a round body and long, narrow neck, intricately carved and decorated, in the central midan at Khartoum. Such is the importance of coffee.) Water in the meantime has been brought to a boil in a tin can on the coals. This is poured into the gebany pot over the crushed beans. The pot is placed on the coals until it begins to make the thickening sound of frothing up into a boil in the neck. It is then removed from the coals with a rag (a shred of an old orange nylon dress, also from one's kit) and placed in the cool sand beside the fire. Three or four tiny china cups are taken out from one's kit, along with a plastic bag of sugar. Sugar is poured into the cups, to fill them half full, then gebany is poured over it and stirred with a matchstick, or twig, and passed to the surrounding company. The cups are filled and refilled and passed until the coffee is gone.

Everything has a use, and is used over and over until it has a certain taste. Fire-blackened things, worn at the edges and one's own through long habit, are best. By the time the gebany is made and drunk, the sky is very dark. The only light is the light of the charcoal of the fire, and the light of the cigarettes glowing around it, and the stars pouring thickly overhead.

Saad Abdullah says to me, "Let me tell you what it was like out here when I was a boy. When the water trucks went by to the phosphate mines, we fell to the ground in fear. We heard the rumbling but didn't know what it was. We used to make fire from a flint, and eat the seeds of desert plants, the wild grass that came up when it rained, if it rained. In those days, in order to survive, you had to know everything. Look at these stars, how lovely they are. We had to know what the risings meant. Every possible indication of season and rain. You wear a watch. But the sky itself is a clock. Here, let me show you."

He begins to draw the constellations with a twig in the fire. He takes me by the arm out into the close darkness and points out Traya, the Pleiades, and the other constellations that belong to the rain, whose setting brings the rain.

The Milky Way, he says, is the Margar el Kabsh, the "path of the sheep." Once people sacrificed children, but then they learned to sacrifice sheep, and God made the Milky Way to commemorate this, the sacrifice of sheep instead of children. It is a covenant.

"Just like the rainbow, gedayagoz," I say, but they have not heard the Old Testament version I know.

A crazy old man comes up, swinging his cane. He rails at Saad Abdullah in a harsh, deep voice, then comes over and peers into my face. As I take the plump hand he offers me, I see that it is Amna, dressed as a man. We all laugh. What a marvelous joke. The joke, I know, is on me, dressed as a man, sitting with the men. Amna takes a bag of hard candy out of her pocket and scatters it over us: the karama.

Amna's house is of multicolored cloth and rugs she made herself, plastered on twisted samra and seyal trunks and branches.

In it she keeps a huge shard of seyal bark from a very old tree, a little brass bell hanging from above, "for haras," protection. We both agree that her house is more pleasant than the plank houses. "It is for the baheim," she says, the herds, the kind of house they used to make up in the desert hills before they came down to camp on the coast. The planking for the seaside shacks is brought from the Nile Valley with the sheets of corrugated tin. And then there is wood that drifts in from the sea. "The sea brings wonderful gifts," Amna says.

Saad and I have just come back from the sea where we went with Saad Abdullah to walk while the tide was out, turning over green crusted heads of coral to find the psychedelic soft living colors of "sea worms": a blob of apricot paste, white flabby tongues, a dense rose-pink growth of sponge. A crinoid, neck craned out, gill petals flowing, dives under pocked coral. Saad wore electric-green plastic shoes and, with his ankle-length underwear pulled up to his upper thighs, looked like a lovely girl swaying beside the solid Saad Abdullah. I had purple plastic shoes on. Saad and I went out to where the waves were breaking on huge colored stones beyond the reef and stood there ecstatically happy. "Il bahr gamil," he said, the sea is beautiful, "agmal min a gebel," even more beautiful than the desert.

When we turned to walk back the water was rushing in around

our feet. I said, "Let's go that way," pointing a long diagonal south to my morning place. "No." He said, "There is a path in both the desert and the sea," and in a moment pointed out a large shark circling there. I trip and take his arm, thrilled slightly with fear. The shark is moving in two or three feet of water. "Matkafeesh," he said. "Don't be afraid. We'll kick it." When we reached shore, a wind was blowing. Not a cold wind or too strong.

He put on a shawl he was carrying and shivering led me to sit against a shalily dune in a scooped-out pit of sand, saying he was freezing, but the sand was beautifully warm. As we walked up to breakfast—the rest of the body of the speckled fish—I told him about snow and fall and how I loved the cold. "The desert and the sea are one," he said, holding two fingers together, "but the desert is more merciful."

"ELBA, ELBA, HAGAR EL ELBA," Saad sang as we slid across the sand dunes south of the Sudan border, fearing no one.

"There is no government down here," he said.

When the Niva boiled over we stopped and made tea, in the long thin shadow of a dune, or under a thorn tree from which a crumbling rope hung down where someone once had slaughtered a gazelle. Saad cleared the ground and set up a fire circle for us, digging up dry roots, and carelessly arranging lovely colored stones.

We traveled through sparse stony wadis filled with samras and dry shoosh, into Wadi Efingab, sandy, with a rare crop of halfbar. In Wadi Aidhab the sand formed low ribbed dunes, greening with plants as we approached the mountain.

We come to a tree, huge for these parts, with a smooth much-folded trunk within dead branches curling down to the ground and living ones growing both up and down—our first indication that we are in a strange place. It is called, Saad says, Kamob. Its branches are flowering now and full of bees. I hear them and think at first it is the sound of the wind. There is a tiny black bird in the leaves with a scarlet throat and blue-green shimmering hood.

Saad says it is the ter al mattar, the bird of the rain, saying "Shall we kill it?" He already has a stone in his hand.

"There may only be one in the world," I say. It is a sunbird of some kind, a shining sunbird. Homoshkeleid is its Beshari name.

"No. There are lots of them here at Elba," Saad says, and throws the stone, saying to Ahmed, meanwhile, "She never wants to kill anything."

As we near the mountain, the trees get thicker and thicker. The ground is covered with sand and cool grey stones. We come upon the humped dun-white mat-covered dwellings of the Besharin. As we stop and climb out of the Niva, two men come from different directions and sit with us. They have a starved, dusty look and eye us suspiciously. We tell them we have come from Port Sudan, the Sudan side, for we have heard that a team of Egyptian geologists were arrested here in the wintertime and are still locked up in Khartoum.

A red-eyed cripple next appeared with a long branch staff,

followed by a toothless Abu Shendab—man with a large moustache—who now sits beside Saad, having accepted a cigarette. A lovely evening descends, thick yellow light, birds in the trees. We are camped where a lagleig and a seyal grow together. A flock, twenty or so, larger than I have ever seen, of turquoise birds—white-throated bee eaters—flies overhead. There are red-breasted doves with strange chattering cries.

With the last sun, quiet, a dream sky, a heavenly rich stirring smell from the trees—all sizes and shapes, like being in a toy shop of trees, a crowd of them. Peach cirrus clouds above a soft blue, swallows, forked-tailed, hundreds of them flying from tree to tree.

I see first one, then another large black-winged rust-bodied bird coming from the mountain. I think they may be lammergeiers. Saad says they are hada, eagles, then nisr, vultures.

He has folded up the Swiss Army knife after preparing tuna fish salad. The two large birds are joined by a dozen others, circling over something in the distance by a pyramidal hill beyond the trees. In the last light they come back toward us and land in the crown of a low lagleig. "Hadeid," Saad says, the final verdict through the field glasses.

I say I want to climb the mountain in the morning and stay up there the following night. "There are leopards up there," he says. "I want to see one," I tell him. "You go right ahead."

A Beshari, black-skinned, slit-eyed (one cannot see his pupils, as though he were blind), in a greenish dun-white shawl and turban the texture of burlap, told us the leopards were eating their sheep. The man came and sat silently while Saad made gaboury, a flat bread baked in the sand under raked coals. We waited for the fast to break. When the Beshari rose to leave, Saad begged him to stay, then spat in disgust when he actually did and ate a good amount. When we asked him, the man said there was no milk.

"No milk! What else do you think they eat up here!" Saad said when he left.

Saad describes the leopard as abraq, speckled. "You know what a cat looks like?" he says.

The ground beneath our seyal tree is covered with lisan a tayr, birds' tongues, dry and red and green.

Ahmed stalks about our circle loudly breaking wood and makes a roaring fire. He sits down with his jar-fashioned water pipe and honey-soaked tobacco. Saad comes out from under his heavy wool green patterned blanket and puts his feet by the fire. I join them. Saad makes me tea with milk, stirred with a thorn twig. I drink it with the sweet chocolate I have been saving for us, just for this moment. We sit shivering gladly by the fire. Saad tells me this is what you do during Ramadan: have a party in the middle of the night, something light, a beautiful, irrelevant gesture, something to get you through the harshness of the following day.

The ground is shimmering with spider webs. We are packing to climb the mountain. After the fire last night I could not get back to sleep. A huge, moist-winged bat swept over me in the dark and I screamed, but neither Saad nor Ahmed heard me.

I rose before dawn, and gathered dead wood from under the thorn trees and tried to start a fire with the bark as tinder. As I was doing this, Saad, who had been sleeping in a heap with Ahmed, for they have only one blanket between them, came over, stretched, picked up a nearby dry root and tossed it on the pile of twigs I was trying to light. He put a match under it.

"Mayinfash," I said, as it caught and died. That doesn't work.

"Yinfa w nuss," he said. It works and a half. He spread his long thin hands over the spark and cupped it as though he were lighting a cigarette. It caught. "Aiz shay walla ay?" You wanted tea, or what?

kundur—dried root of shoosh

Beshari elders follow us and say that if we near the cave they will break our legs with their walking sticks. They hold them stiffly over their shoulders, and mean it.

We stop at a well on the slope of Elba. There are eagles circling above, large, eagle-like birds, melaun, as Saad describes them, mottled: back white, belly red, wings black above and white below, head black. Lammergeiers.

We meet a pleasant young Beshari named Hassan at the well. He tells us about the moss-covered cave within the mountain that is said to hold the spirit of the mountain, its breath. Hassan says you never see the leopards, but they are here. You see their tracks. They come to eat the sheep, but never Beni Adam. He pokes the outline of a leopard track in the sand.

"Not true," says Saad. "They eat anything."

We climb Akau, the gorge above Hassan's home, where there is a spring, and above it "the eagles' cliff." A man is watering his sheep at a stone trough. Saad is showing me how to scrape a yellow flowering iris-like plant with soft hairs across my tongue. It has a tart lemon taste. He shows me sakhran, "drunk," a poisonous plant, that in small amounts is a mild drug.

Hassan runs barefoot up a precipice to rescue a goat from a high narrow cliff. Fig trees grow among the steep grey rocks. Their figs, Hassan tells us, are especially delicious, and give you remarkable strength. The fig trees have long pointed leaves, like blue-green gums, but are dark and shiny. There are smooth, colored rocks below with a coolness like water where many plants grow, tiny flowering plants and vines. Giant seyal and other trees and their delicious shade.

Saad and I climb along the cliffs. Hawks and eagles fly close beside us. "Hatawga," we chant singsong to each other—You're going to fall. We each answer: "Matkafsh, yachy"—Don't bet on it, brother—"Matkafeesh"—Don't you fear.

We descend to a deep rock pool, where a stream falls thirty feet down a mossy rock face. We decide to climb down the almost smooth rock, despite the shouted warnings of the Beshari who are watering their animals at a pool below.

Saad scrambles in front of me to the ground, then climbs up to catch me as I come to where there is no further foothold. I stand against the wet, mossy rock, then descend rapidly into him. We remain there precariously plastered for a minute, trembling, his foot on mine, then jump the short distance to the stony sand below.

This feat has everyone laughing. The children bring me plants, the knife-stiff fronds of dragon's blood trees, from an upper slope. My arms are scraped. My clothes are wet. My back hurts. I am happy. Honey bees are drinking the water from the rock wall. Clumps of tall pink and white hollyhocks grow against the boulders.

Our three pairs of slender, similar ankles are stretched out, Hassan's, Saad's, and mine (the thickest), black, brown, white.

"The sea and the desert are one." Hassan says, "There is danger in them both. They both break you." The three of us are lying among the rocks beneath a thorn tree, splitting an orange. I am exhausted. "When you leave the sea, you leave it. When you come down from a mountain, it stays in you," Saad says.

We are on a windswept plain at the opening of Wadi Hodein. You go to Wadi Allagi this way, about a hundred eighty kilometers in over the desert from here. The samras are bent slightly back, as though formed by the wind. We are ten kilometers from the sea. The sand is white with quartz. It is, Saad says, naama, light, as opposed to kashin-gamid, hard. The sun has suddenly lit the plain, rising beyond strips of cloud.

There is nothing growing but samra here. Under the sand are clumps of dry roots. Near us is a bone scatter, the bones of one or more large animals scattered white and beautiful, a brilliant white on the sand.

After coming down from Elba I was suddenly exhausted and could hardly walk. Hassan made me a walking stick of dry wood, "like a third leg." We went down slowly; it took an hour or so. At the bottom, I collapsed on a rapidly provided rug beneath a thorn tree, shivering with fever, wrapping myself in warm clothes.

After two hours or so, the fever started going down—"Just

like the car," Saad said—and my appetite came back, but I was still weak.

As the light thickens into evening, we say to each other, as we often do, "Nabeit fayn?" Where shall we make our house? Beit, beta, the letter B, the first architectural plan. We stopped here in a strong wind. A large hare, fawn-colored with very white feet, ran out over the sand from beneath a samra tree. Saad ran after it on foot, but he didn't catch it. Now, waiting for me to finish with my notes, they have dug a checkerboard in the sand and are playing siga-shararank with stones and camel dung. A game that they have been playing here for thousands of years.

In Shallatein, the border town, the people are "Bisharin," but they look a mix of West African and Beja and, as Saad says, aya kalam, who knows what. The houses are made of crates, plywood, and (some with quite solid) planking. All have front porches and pointed roofs. It is a wonderful kind of architecture, made of any kind of material—the plan exists, the pointed roofs, the porches and corals, and they fit scraps of wood and tin to it any which way. It is also remarkably clean.

Saad and Ahmed have just walked off, leaving me in the car, saying, "Tiktib w ma kalemsh ayahed"—"Write, and don't talk to anybody. Tell them we've come from Allagi. Better yet, you don't speak Arabic at all, okay?" The car is filling with flies.

—

The road here, the main street in Shallatein, is spotted with flattened tin cans. A bird with a blue throat and a scarlet bib is near the car, a cordon-bleu. Last night Saad and Ahmed dug a hollow in the sand to sleep in, having been terribly cold the night before. "We couldn't sleep at Elba," we were all saying. "So damp. So cold." After dawn a white vulture came up from the sea, its wingtips black as though dipped in ink, and circled my little samra tree, followed by two ravens.

Saad told us that he was born here, in this wadi near Shallatein.

"Were you happy here?" I ask.

"Of course. I was a child. I didn't understand anything."

A STRONG NORTH WIND often blows on the Egyptian side of the Red Sea. Though some parts of the shore are clear—at Abu Ghusun for instance, where Saad Abdullah is—where the coast juts out even slightly, the garbage is knee-deep along the tide line. On the north side of the large peninsula Ras Benas there was once a small military airport. Tangled barbed wire closes off a ground visibly strewn with land mines. The peninsula swings far out into the sea here, and its outer edge is a treasure trove of odd objects from around the world.

The most valuable things that drift in are wood planks, which can be used to build houses, and rope. There was one coil of heavy blue plastic rope half sunk in the sand that everyone tried to dig out but failed. Then there are aerosol cans which serve as toys for the children, and any number of mysterious plastic things—but watch out, for one could be a lagham, a mine or "anti-personnel" bomb, which are found not infrequently. Two men were blown up handling such a mine near Abu Ghusun and, legless, bled to death on the way to the airlift at Berenice. There are many potentially useful household items beneath a heavy persistent covering of tattered black plastic.

One day Abdel Hadi and I found an octagonal blue glass bottle still with perfume in it (each of us wanted it for our afesh because of its color, but were both too shy to say so), and a long wood spool that made a perfect rolling pin for ragag. Abdel Hadi's father, Ahmed Ali Kheir, the withered black fisherman with two wives (both twice as big as he), found an expensive Japanese underwater camera in his stretch of mangroves. But no one could figure out how to open it, much less use it, and it sits in his seaside shack, a mere suggestion of another world.

IN LATE SPRING I went to Cairo for a break. The city had become a place of intoxicating luxuries—silks, shadows, water. I stand in Anna's room of silks—Anna, the Italian fashion model who married David's friend Zeitun and bought a silk factory in Alexandria.

I look at myself in her wall-size mirror: wrists, hands, and face thin and brown, body ghostly, almost greenish-white. I pick up swaths of silk and wrap their soft, delightful texture, their beautiful colors, lime green, plum rose, against my skin. I watch, all the while, the strangeness of my face, seeing in it (for I have not seen it in a mirror for a long time) my mother, my father, my brother Charlie, my lost brother and sister: not simply the structure of their faces, but the very look in their eyes.

I will be in America soon, surrounded by familiar things, and this rare world of Kharit, of Hamata and Elba and Abu Ghusun, will recede and begin to seem like a dream.

David asks have I not heard stories about the insecticide dumping in the Elba area? No one knows whether they are true or not—whether there was mass spraying in the desert to stop the plague of locusts last year. Hadn't they come from the greening of the desert around Elba after the rain? Is anything left alive out there now? Well, anyway, he loves the word for insecticide: mubeed hesharat.

David has fallen in love. He has more than fallen in love. He has become domestic. A grey-eyed Canadian with dark hair and pale skin wants him in the kitchen, right now. How lonely I suddenly feel.

I drive into the desert toward Wadi Kharit from Kom Ombo as the sun sinks into a brown haze. A sandstorm comes up, thick as a blizzard. Farther on I can see lightning striping it, heavy and low, frightening, too close. Rain pours down in sheets.

My gas gauge shows empty. All day I have been afraid that my worn tires will pop. The road is impossibly bad. The windshield wipers do not work.

In Wadi Kharit Ababda sit out in the dangerous lightning, as if watching a delightful show.

Hasabullah, Saad's little brother, runs out to meet me as the car dies, just as I reach the edge of their camp. Saad is racing around in the rain, herding the women and children into the rooms of the empty mudbrick structure that stands among their domed branch-frame huts. He searches frantically, delightedly, for a lamp, and finally makes one out of a cough syrup bottle and some rope.

We cluster in the hot damp room. I sit against the wall, nodding off to listen to the wonderful sound of the rain, between the pregnant Asha, Saad's wife, and her pregnant sister, Saad Abdullah's wife. The water streams down through the unfinished roof as the rain beats it in, dissolving it. Saad hands me the lamp to help lead the migration out into the rain again.

The sky is bright with blue, lavender, pink flashes of light.

"So what do you think of the rain?" We scream at each other.

"Azim." Fantastic.

At the geological survey station in Marsa Alam, I have just come up from the beach, where the director, Mr. Rifaat, and I were sitting in the dark against the hull of an overturned boat. We were singing to each other, songs that brought tears to our eyes.

"Let us walk on the beach and sing moving songs," he had said after we finished our supper of tomatoes and cheese.

Mr. Rifaat is an Alexandrian, a sad, droopy, light-skinned man in his sixties. I thought of the Walrus and the Carpenter as we walked along.

We tried to make each other cry. He sang Abdel Hafez Halim. "I traveled with him, I traveled with him. O God, if only you would give me another lifetime to do it again."

I thought of Gamal and Joe, and Saad and Abdullah, and how intensely I would miss them when I was back in America. I thought of something Saad once said to me, as a joke, "Out here, people are your furniture, your house."

I was about to sing Waylon Jennings's "Wurlitzer Prize," but Mr. Rifaat wanted to hear Shirley Bassey's "Never, Never, Never." I did my best with it. I hadn't heard it since high school.

In the morning I woke to a pounding on the door. Saad Abdullah had come up from Abu Ghusun in the night, having heard I was back in these parts. He brought me a breakfast of unicorn fish.

A tall heron is standing on the reef in the breaking surf. The sun has just risen, at first a ghost-thin wafer, then a cloud of warm lemon-green light. In the waves before the sun move the undulating bodies of two large sea animals, dolphins or sharks, their tails and fins and heads appearing and sinking and rising again in the waves they ride. A quiet sea today.

Saad Abdullah said to me last night, "You need a change of

weather, Doctura, rest, good food. You're really getting worn out here. You look it." This was partly to annoy his mother who was putting together lentils to pour on the bread the children and Saad were shredding into the communal bowl. They had gone through their usual bantering of, "What on earth are we going to eat? There's nothing here." This, I think, is a medium of their fondness for each other. They are two very tough people. It is also partly true.

I said, "Oh, shut up," and put my hand around his large weather-worn foot, deeply cracked by the heat and sand and dryness, toes bent and squarish like piano keys.

He jerked his head back into the nod and smile of our familiarity, silver teeth shining in the dark.

We could just barely see each other.

We could hear Fatna singing and beating the tar out over the sand.

Later I went to watch her dance in her princess rags—a scarlet synthetic silk from Shallatein, torn and dirty. She moved jerkily, untaught, her twelve-year-old body full of a wild, almost angry pride, and her face, as always, hard and breathtakingly beautiful.

The little Greyjab girl with the scarred face—as though a shallow, broad circle of flesh had been scraped out of the right side of her face—had learned to dance in Edfu. Picking up the corners of her yellow gown, she skipped around our circle in a slow, elegant way, one back heel lifted like a pedal to push the hip rhythmically up, her head thrown back.

Saad Abdullah came back from Shallatein over the weekend with Fatna, whose father, a distant relative, had drowned recently when his boat capsized on a trip to Halaib. Her people are south coast fishermen, with fine long bones and delicate features and lavender-green eyes. He needed a girl to help Amna cook and clean in the camp. But more than this—he had been an orphan, fatherless, himself, and is known for taking in orphans now that he is a man. Her little deaf-mute brother came with her. The other children call him atrash, dummy, and hit him because he will not cry.

Later, as I lay sleepless in Saad Abdullah's mandara, his "guest house," under my sleeping bag, slightly feverish, and frightened of illness as the people here are, a swollen gland in my groin, two oozing skin infections along my left jaw and one, drawing flies, on my right ankle, two Beshari traders arrived on their way north from the border.

Saad Abdullah lay cradling several children on the sand floor of the next room.

The dim yellow light of the kerosene lamp came through the thin wall.

I heard the loud, unfamiliar voices of the Besharin. Then

Saad's voice, sweet, throaty, in a higher register—had they had dinner? Could he get them some tea? When he came back from Amna, roused to get this tea for them, they said they had heard there was a hawagia, a foreigner, living here. Where was she?

At first he ignored the question. Then he said, "Look, it's late. We're tired. I've gotten up to get you tea. Let's go to sleep."

There is a very pleasing sense of belonging now, of being under Saad Abdullah's protection, along with the others in the camp.

A small silver fish skips like a stone over the surface of the water, chased to its death, a black swirling underneath. A powerful wind has just come up.

A delicious lunch of sigan fish and their roe with Abdel Hadi, who caught them out fishing with his mother at dawn. He is telling me about the kukar, what they call the draysha here, the horned viper. Once he saw one in the shadow of a clump of camel thorn and put a long acacia spike through its head—not killing it—then wrapped a string around its mouth. It lived for another four hours. He says in a day he might see five or six in the house. But "hetta hetta yaod," they only bite now and then, and it is pointless to fear them. His mother says they sting the tracks you leave in the sand and that even their bones are poisonous. Abdel Hadi, at fifteen, is the better naturalist.

As we walk in the mangroves when the tide is out, he disappears into a tree and brings out a dry whitish snakeskin, the garment of the snake. He says he has seen it there for a month and a half.

A purple heron with a white spot on each glossy wing dances in the shallow water among the trees, swallowing small crabs and silver fish with its long curved yellow-orange beak, long dangling feet the same color. It has a hairlike crest.

The coastline abruptly hooks north, and the tattered plastic along the shore is knee-deep. Translucent, brightly colored bags fill with wind and move inland, like creatures emerging from the sea. They drift up the mouths of the wadis and travel for miles, sticking to the thorns of desert scrub and trees. "The desert eats everything," Abdel Hadi observes.

yelalag—I have been waiting almost twenty years to hear this word. Someone said to me last night, "The foxes babble with joy in the hills when it rains."

I know the word from a poem I once learned from Steele Commager:

> Pone sub curru nimium propinqui
> Solis in terra domibus negata
> Dulce ridentem Lalagen amabo,
> Dulce loquentem

Put me beneath the chariot of the too near sun
in a land denied even of houses
Still will I love my sweetly laughing Lalage,
my Lalage sweetly speaking

Lalage, the sound all of nature makes, the language of birds.

It is dark and late. I am writing under the thick stars by lantern light.

We are silently waiting for Saad to come back, watching for the eerie glow of car lights beyond the sea hills miles away.

Hassan Karrar is singing softly, sweetly, as he fetches trays and bowls and fried fish strung up on leather straps in a string bag in our kheima for supper.

"Il dunya helwa, helwa," he sings. The world is beautiful, beautiful. Beyond this the insistent waves.

EPILOGUE

MY HUSBAND AND I used to say that we got married because we knew the same poetic line. The line was from John Shade's poem in Nabokov's *Pale Fire.* The first time we had dinner together, we were talking about the emotional impact of color. My husband said, as an aside, as though he were simply remembering the words to himself, "All colors made me happy: even gray," and I at once came back with the other half of the couplet: "My eyes were such that literally they took photographs." We both secretly prized the line, and were surprised that the other would know it, as if we both knew the same lost language.

It is a long time since I came back from Egypt. Lanny almost died a few years ago. We live in the country. We are in the mountains, and fall comes early. Now it is settling in over the sugar maples streaking them with peach and red. Up on the hill the frogs in the pond are almost as big as my hand. They are brown, with green faces and golden eyes. The deer come down from the woods at night and eat the apples off the trees. Home.

Reeds of Re

- i. leaves of reeds

- ius - reeds

- tmw - worm? (Rec. B131)

- ihmw-wrd - the unwearying stars

- indestructible star (circumpolar) Pyr. 782

- isr - Tamarisk

- isd - unknown tree Pyr. 95 (has fruit)

- sshmw - a species of bat

- a Milk Goddess Pyr 131

- isrw - Rushes

- int - valley

fish and decay - be abomination

- ss - cedar

- ssn - lily

- bulls head on a pole

- indt - Nile Acacia (Sont)

- sb3 - star

- Tnmw - darkness, gloom

- hpš - (Orion) Ursa Major

- sh - Orion

φρισσω - Bristle freeze stiff

δαρ - a wife

φρενιτιάω - to have a violent fever, be frantic

φρατηρ - member of a clan

ACKNOWLEDGMENTS

In Egypt the best thing you can say about a person is that he is generous. This book has evolved through the extraordinary generosity of a number of people.

I would first like to thank the great desert naturalist J. J. Hobbs, who freely shared his insights and his resources, and gave much-needed encouragement through dark times. He is well known for his integrity, and his name alone can purchase food and shelter almost anywhere in the Eastern Desert and Sinai. I also would like to express my deepest gratitude to Gamal Sharif, who has been an invaluable friend and helpmate for a dozen years.

I hope with this book to contradict the notion that Western women are somehow at risk in the Middle East. Indeed, they are normally treated with unusual kindness and solicitude. Niazi

Taher went to great lengths to help me secure needed permissions, simply out of friendship. Dr. Nabil el Hadidi gave me the sponsorship of Cairo University and the Cairo Herbarium, and helped me identify plant environments, and plant specimens from the Elba area. The monks at the monasteries of St. Paul and St. Anthony have taken me in, fed me, allowed me to attend their ceremonies, and long given me good company and any needed help on my journeys north and south. I would particularly like to thank Father Dioscuros, the librarian at St. Anthony's, and the brilliant icon painter Father Bishoi, and Father Agathon at St. Paul's. The Ababda of Wadi Kharit, Abu Ghusun, and Hamata—Saad Mohammed Ali Kheir and his cousin Saad Abdullah and their families, and the family of Abdullah ibn Saad of the Kreyjab—not only welcomed me into their lives, but were wonderful friends and companions, and I miss them acutely to this day. Dr. Said Hurreiz at the University of Khartoum helped me considerably in the difficult spring of 1989. A conversation with the great Dinka poet Francis Deng convinced me that even the writing of an outsider on the poignant subject of Sudan was of value.

The Crane-Rogers Foundation funded my research in Egypt and Sudan from 1988 to 1990. My special thanks to the foundation's director, Peter Bird Martin. Bob Betts and the staff of the American Research Center in Egypt helped with the various problems of living in Cairo. David Loggi at the Pierpont Morgan Library provided the photographs of my journals used as illustrations in this book.

I would like to thank many friends for their great kindness, and for conversations that have helped to form this book, especially David Sims, Elizabeth Wickett, Michael and Angela Jones, Carl Strehlke, Dickinson Miller, Mark Lehner, Lillette el Biali, Lilian Karnouk, Milbry Polk, Sally Wriggins, Jeanette Watson, Jonathan Rabinowitz, Rob Cowley, and Fred Grunfeld. Special thanks to Leila Hadley Luce for much-needed advice and support. My very dear friend the Venerable Khyongla Rato Rinpoche gave invaluable insights and encouragement.

My gratitude to my editor, Amy Hertz, and my agent at Janklow and Nesbit, Cynthia Cannell, and to the staff of Riverhead Books for their beautiful job on the production of this book.

Last of all I would like to thank my parents, David and Shirley Brind, for their enduring love and constant help; my brother Charles and my great-aunt Dorothy Murray Sliter, who have always been an inspiration to me; and, for his love, his patience, and his unmatched skill with the English language, my husband, Lance Morrow.

AUTHOR'S NOTE

There is nothing in this book that did not actually take place. I have occasionally compressed the events of years into a single episode.

All the translations in the book are mine, except where otherwise specified (Robert Lowell's translation of the *Oresteia*), as are the ideas about language and myth. Apart from translations, I have quoted Cole Porter, "I Get a Kick Out of You"; W. B. Yeats, "Lines Written in Dejection"; Steven Vincent Bénet, "American Names"; Alcman, fragment 26; Horace, first book of *Odes*, 22, 37, 38; and the Quran, suras 91, 105, 113.

Instead of giving an account of the natural history of the Eastern Desert, I have tried to present animals, birds, stars, and so on as people described them to me, using their voices. Hence the bissa is the hawksbill turtle; the arussat al baher, the bride of the sea, is the dugong; abu salama, the servant of peace, is the dolphin; abu grun, the unicorn fish, is the triggerfish; and so on. The lynx mentioned is the itf, or "swamp lynx," and is a different creature entirely from the Canada lynx. The information on the number of flamingos on Lake Bardawil is from the Egyptian Conservation Department in the spring of 1990.